Concepts of classical mechanics

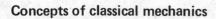

European Physics Series

Consulting Editor

Professor J. M. Cassels, M.A., Ph.D., F.R.S.

Lyon Jones Professor of Physics
University of Liverpool

The lunar module, attached to the Saturn V third stage, photographed from the command module. The photograph was taken during the Apollo 9 flight, preparatory to the first moon landing, which was accomplished by Apollo 11.

(Associated Press)

Concepts of
Classical Mechanics

L. R. B. Elton

Professor of Science Education
and formerly Professor of Physics
University of Surrey

'I could be bounded in a nutshell and
count myself a king of infinite space.'

Shakespeare

McGRAW-HILL

New York · St Louis · San Francisco · Düsseldorf · Johannesburg
Kuala Lumpur · London · Mexico · Montreal · New Delhi · Panama
Rio de Janeiro · Singapore · Sydney · Toronto

Published by

McGRAW-HILL Publishing Company Limited

MAIDENHEAD · BERKSHIRE · ENGLAND

07 094159 9

PRINTED AND BOUND IN GREAT BRITAIN

Contents

To my wife

Preface

The subject of mechanics occupies a very large and probably disproportionate part of the average school physics course. As a consequence, students are apt to enter their first mechanics lecture at university with a feeling that they have been there before—several times. In fact many are quite adept at handling problems of considerable complexity, but have never been stimulated to reflect on, or taught to understand, the fundamental basis of what is, after all, one of the supreme creations of the human mind. This situation leads to interesting problems in the presentation of the subject, and the present volume is an attempt to solve these at a level at which even moderate students can take to it and good students will be stimulated to further study in the field. The aim of the book is to give students an understanding of the basis of the subject and of its historical importance, and to indicate some modern applications. It is not designed to give them dexterity in the solution of standard problems in applied mathematics. Mathematical proofs and derivations are included for the sake of better understanding, but students should not be expected to remember details of these or to have to reproduce them under examination conditions.

The book has had the benefit of an exceptional amount of field trials and feedback. It is based on a lecture course, given for the past seven years to a common first-year group at the University of Surrey, containing all the students in the honours courses of physics, physical science, metallurgy and electrical engineering. The students were given very full printed lecture notes, which were revised annually in the light of experience and eventually led to this book, although this contains more material than was presented in the lectures in any one year. The choice of subject matter has been affected to some extent by parallel courses, given to the same students, but far more important has been the influence that the nature of the group has had on the treatment of the subject. In many universities, physics departments offer two kinds of physics courses, for physicists and for non-physicists, and the latter frequently do not get the same attention from the department as the former. In other universities, non-physicists attend courses designed for physicists, and this too is not always satisfactory. The present lecture course, given equally for four departments, has been hammered into shape in long discussions between the departments, followed by trials, further discussions and further trials. The resulting book attempts to give a broad view of the subject, equally suitable and stimulating for mathematicians, physicists and engineers, and of interest also to chemists and metallurgists. At the same time, it has been designed as a suitable preparation for more advanced courses in the field, whether taken by physicists or mathematicians, and in view of the considerations presented at the beginning of this preface, it is hoped that it may also be of value to schoolteachers, who may treat much of it as 'elementary physics from an advanced point of view'.

When the manuscript had reached draft stage, it was used in a discussion group of staff, students and schoolteachers in mathematics and physics, who met weekly for a term. Their forthright comments on major matters and meticulous attention to detail has led to many changes and improvements, although it would be too much to hope that it has led to the elimination of all errors, for which I must remain solely responsible. The members of the group, to whom I wish to express my sincere gratitude, were Mr R. G. H. Bloomfield, Mr O. M. Dashwood, Mr D. Eden, Mr D. D. Gibson, Mr R. D. Harris, Mr H. G. Hibbert, Mr D. L. Hurd, Miss Mildred Piper, Mr F. D. Tyler and Mrs Jill Harvey. In addition, I am most grateful to Dr N. Ream, who read and commented on the whole book, and to Dr A. W. Wilson, for help with the flow chart on p. xvi. The book also owes much to the approximately 1000 students who attended my lectures, and I wish to thank them for their helpfulness. Finally, I wish to thank Mrs Daphne Whorlow and Mrs Jenni Thackwray for their work in preparing successive drafts of the manuscript.

L. R. B. ELTON

University of Surrey
1971

How to use this book— a note to the reader

The book has been written explicitly for self-study, whether in conjunction with a course of lectures or not. This approach arose out of an investigation carried out at the University of Surrey, which showed that students often find it very difficult to organise their study time, when they first come up to University. Several of the features of the book attempt to meet this point, the most important being the self-tests described below. Figures are arranged to illustrate points in the text immediately adjacent to them, in a way in which a lecturer illustrates his talk with rapid sketches on the blackboard. Similarly, I have incorporated asides, which I have found illuminating in my lectures, even if at times they interrupt the flow of an argument. The book should therefore be treated as a written set of lectures, and you should have pencil and paper at hand in order to elaborate, fill in gaps, write out derivations in full, etc. In this way, between us we shall write a text book! If, at first, the material seems too much, then some or all of chapters 4 and 7, as well as sections 2.12, 3.7, 3.8, 6.6-6.8 may be omitted.

I have kept references down to the minimum, in the hope that you will follow them up and read the books and articles mentioned, and for the same reason have given the references in the text. I would like to draw particular attention to the following:

1. M. Born, *Einstein's Theory of Relativity*, Dover
   This is a classic by one of the great masters, which has lost nothing in the fifty years since it was written, originally to introduce the lay public to the then new ideas of Einstein. Since it was written for a lay public, the new ideas were prefaced by an account of the old— in mechanics, optics and electromagnetism. To this day, it serves as an adult introduction to the ideas old and new, of physics. It forms excellent background reading for the present volume.

2. Daphne F. Jackson, *Concepts of Atomic Physics,* McGraw-Hill
   Many of the applications of classical mechanics occur in atomic and nuclear physics, and I have mentioned a number of these. Of the

many textbooks on the subject, I have chosen Professor Jackson's book, which has grown out of a lecture course given to the same students who also took my course, and I have given references where appropriate.

3. Articles from the journal *Scientific American*
   Over the years, *Scientific American*, with its combination of clarity and authority, has become deservedly famous. I have selected relevant articles, and quoted them by author, title, offprint number and reference. An order form for offprints can be obtained from W. H. Freeman & Co. Ltd., Warner House, Folkestone, Kent. (The journal is abbreviated to *SA* in the book.)

4. Other references
   These are in general on particular points related to the text. I have not given any advice on what might be called 'further reading', since the field here is embarrassingly large and varied, and what might suit one, would not suit another. A few hours' intelligent browsing through the mechanics section of any good library will serve you much better.

5. Lastly, I should like to mention two excellent books which illustrate just how much of physics—and not only mechanics—is illustrated in simple everyday things. They are:

   Elizabeth A. Wood, *Science for the Airplane Passenger*, Houghton, Mifflin.
   R. A. R. Tricker and B. J. K. Tricker, *The Science of Movement*, Mills and Boon.

   There are four kinds of examples:

(a) Progress tests consisting of multiple-choice questions. These enable you to test for yourself whether you have understood the subject matter up to the point at which the test occurs. You are strongly advised to tackle each test as you come to it along the lines of the flow diagram on p. xvi, and also, in each case, not only to find the right response, but to study the wrong ones and make sure that you understand why they are wrong. This is a good way to reinforce your knowledge and understanding of the subject.

(b) Worked examples. It is a good idea to have a go at these before reading the solutions.

(c) Problems at the end of each chapter. Do them, because nobody can get a real grasp of mechanics, unless he solves problems in it. Some of the problems are for practice, while others extend your knowledge. Difficult ones are marked with an asterisk.

(d) A final multiple-choice revision test.

The knowledge which I assume that you possess at the beginning of the book consists of some elementary calculus and co-ordinate geometry, and an acquaintance with basic mechanics. All this is substantially less than is contained in standard A-level courses in mathematics and physics, but more than nothing. The calculus is really essential, and if you find yourself inadequate in it, you might find help in a programmed text, such as D. Kleppner and N. Ramsey, *Quick Calculus*, Wiley. A previous exposure to elementary vector algebra is helpful, otherwise the book should be accompanied by a parallel book in vector algebra. You may like the programmed text by K. L. Gardner, *A Programmed Vector Algebra*, Oxford University Press, although this will teach you more than you need for this book. Another useful programmed text to accompany the study of this book is J. A. Taylor, *Programmed Study Aid for Introductory Physics, Part 1: Mechanics*, Addison-Wesley.

As regards units, I have used the international system (SI) and followed the recommendations of the report *SI units signs, symbols and abbreviations* of the Education (Research) Committee of the Association for Science Education.

Finally, I enjoyed writing this book; I hope that you will enjoy reading it.

Procedure for tackling progress test questions

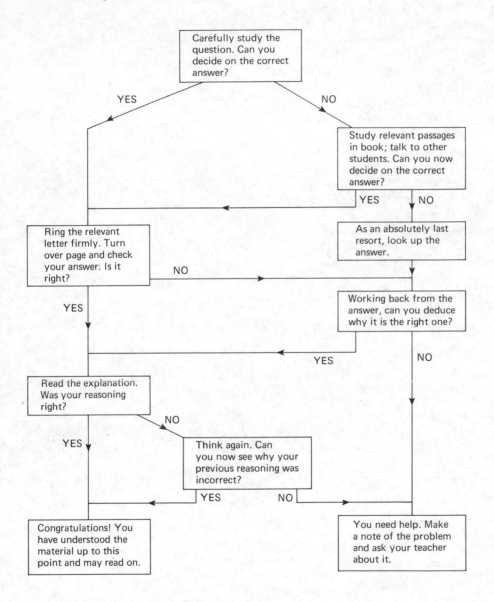

Carefully study the question. Can you decide on the correct answer?

YES

NO

Study relevant passages in book; talk to other students. Can you now decide on the correct answer?

YES

NO

Ring the relevant letter firmly. Turn over page and check your answer: Is it right?

As an absolutely last resort, look up the answer.

NO

YES

Working back from the answer, can you deduce why it is the right one?

YES

NO

Read the explanation. Was your reasoning right?

NO

YES

Think again. Can you now see why your previous reasoning was incorrect?

YES

NO

Congratulations! You have understood the material up to this point and may read on.

You need help. Make a note of the problem and ask your teacher about it.

The Greek alphabet

A	α	alpha		N	ν	nu
B	β	beta		Ξ	ξ	xi
Γ	γ	gamma		O	o	omicron
Δ	δ	delta		Π	π	pi
E	ϵ	epsilon		P	ρ	rho
Z	ζ	zeta		Σ	σ	sigma
H	η	eta		T	τ	tau
Θ	θ	theta		Υ	υ	upsilon
I	ι	iota		Φ	ϕ, φ	phi
K	κ	kappa		X	χ	chi
Λ	λ	lambda		Ψ	ψ	psi
M	μ	mu		Ω	ω	omega

1. The laws of mechanics

1.1 How it began

When a stationary body is given a sufficiently hard push, it starts moving, and unless the push is maintained, the motion sooner or later stops. This elementary observational fact formed the basis of the study of motion of the Greek philosopher Aristotle (384–322 BC), and it led him to postulate that an agent was needed in order to start and to maintain motion. Even though our observations of motion now extend vastly beyond those available to Aristotle, his conclusion is still in accord with experience.

The trouble with Aristotle's conclusion is that it did not lead anywhere, and although the reason for this is now obvious to us, it took the genius of Galileo (1564–1642) to see it in the first instance. Let us see how he argues the matter* in his book *Dialogues Concerning the Two Chief World Systems*, in which he is represented by Salviati, Simplicio is an Aristotelian and Sagredo acts as umpire:

SALVIATI: Now tell me what would happen to the same movable body placed upon a surface with no slope upward or downward.

SIMPLICIO: Here I must think a moment about my reply. There being no downward slope, there can be no natural tendency toward motion; and there being no upward slope, there can be no resistance to being moved, so there would be indifference between the propensity and the resistance to motion. Therefore it seems to me that it ought naturally to remain stable. But I forgot; it was not so very long ago that Sagredo gave me to understand that this is what would happen.

SALVIATI: I believe it would do so if one set the ball down firmly. But what would happen if it were given an impetus in any direction?

SIMPLICIO: It must follow that it would move in that direction.

SALVIATI: But with what sort of movement? One continually accelerated, as on the downward plane, or increasingly retarded as on the upward one?

* The passage is quoted in Ingard and Kraushaar, *Introduction to Mechanics and Waves*, Addison-Wesley, 1960.

SIMPLICIO: I cannot see any cause for acceleration or deceleration, there being no slope upward or downward.

SALVIATI: Exactly so. But if there is no cause for the ball's retardation, there ought to be still less for its coming to rest; so how far would you have the ball continue to move?

SIMPLICIO: As far as the extension of the surface continued without rising or falling.

SALVIATI: Then if such a space were unbounded, the motion on it would likewise be boundless? That is, perpetual?

SIMPLICIO: It seems so to me, if the movable body were of durable material.

SALVIATI: That is of course assumed, since we said that all external and accidental impediments were to be removed, and any fragility on the part of the moving body would in this case be one of the accidental impediments.

Note that Galileo here describes an *idealized experiment*, which cannot be performed in practice. It is only in our minds that we can remove 'all external and accidental impediments'. He realized that only by divorcing the study of the motion from the study of the agents—which we now call forces—was it possible to make progress, and the method by which he did this—abstraction from experiment—remains the basic and most powerful tool of physics today. So at the very beginning of our science we see that the study of physics involves more than observation and experience. It involves experiment, which is deliberate and organized experience, followed by theoretical considerations.

The idealized experiment is also the first stage towards the formulation of a mathematical model of the real situation. Thus, a massive body may first of all be idealized into a point mass, and this in turn may be represented by a number, giving its mass, and three numbers giving the co-ordinates of the point. The translation of a real situation into a mathematical model is often the hardest part of solving a problem in science, since it is an inductive process; the subsequent mathematical treatment and the re-translation of the results into the real situation are deductive, and to that extent more predictable. If anything deserves the title of 'the scientific method', then it is the process of mathematical model making. Without it, science can make little progress.

One of the difficulties in studying mechanics is that the basic model building took place so long ago that we tend to forget about it and think that we are dealing directly with reality. This is not so, and the whole of the rest of this book deals with the mathematical model that has been constructed over the centuries, in order that we may understand better the real motion of real bodies. It may be noted that almost the opposite difficulty arises in atomic physics, where the models that we construct are both new and unfamiliar. [See Jackson, chapter 1.] However, the basic method through which we study natural phenomena is the same in both fields.

Progress test

1.   I have
     A   read the section 'How to use this book' and understood it
     B   read it, but have not understood it
     C   not read it, but know where to find it
     D   not yet found it.

1.2   Length and time interval

Physics was the first science to be truly quantitative, and it was the association of observation with numbers, which we call measurement, that made it possible to apply the powerful tool of the logic of mathematics to physical processes. For a long time it was thought that physical concepts could be defined independently of the concepts involved in observation and measurement, but the work of the last sixty years, particularly that of Einstein (1879–1955), has shown that this is a dangerous practice, which can lead to serious error [R. Fürth, 'The Limits of Measurement', *SA* 255, July 1950, p. 48]. Since experiment and measurement are primary in physics, it is right and proper that everything else pertaining to our study must derive from these, and we must not feed into our system items which derive from, say, purely philosophical considerations. We shall therefore at this stage refrain from considering such philosophical concepts as space and time, and instead define operationally the concepts of length and time interval through measurement. This in turn is defined as a numerical comparison with a standard.

Until recently, the standards of length and time were based on the length of a certain platinum bar and on the time taken by the earth in its orbit round the sun. Modern standards are based on the wavelengths and frequencies of vibration of certain spectral lines. [A. V. Astin, 'Standards of Measurement', *SA* 326, June 1968, p. 50]. They can be understood fully only from a study of atomic physics [see Jackson, chapters 5, 6, 8; H. Lyons, 'Atomic Clocks', *SA* 225, February 1957, p. 71], but it is sufficient for our purposes to know that they exist and that any length and time interval can be measured in terms of them.

The standard of length is the metre, which is defined in terms of the wavelength λ of the orange spectral line emitted by an atom of the krypton isotope ^{86}Kr, through

1 metre (m) = 1 650 763·73 λ.

[H. Barrett, *Contemporary Physics*, 3, 415 (1962).] The standard of time interval is the second, which is defined in terms of the frequency f

4

[1.3

of the spectral line corresponding to the transition between the two
hyperfine levels of the ground state of the atom of the caesium isotope
^{133}Ce, through

$$1 \text{ second (s)} = 9\ 192\ 631\ 770\ f^{-1}.$$

[*Physics Today*, August 1968, p. 60.]

Having defined length and time interval, we can now define
position of a point and *time of an event*. The former is the length (or
distance) between the point P and an arbitrary reference point O, where
we at present confine ourselves to one dimension. The latter is the time
interval between the occurrence of the event and of an arbitrary reference event, which is generally taken to be a particular pointer reading
of the hands of a standard clock, say when both hands point to the
figure 12 on the dial. Thus, from the point of view of actual use, our
concepts of both space and time are essentially relative.

1.3   Kinematics, the study of motion

We have already learnt from Galileo the importance of separating the
study of motion from the study of what causes motion. The former is
called kinematics, the latter dynamics. In kinematics we are concerned
with the observation of the position of a body at different times, and—
with Galileo—we shall consider mainly bodies that can be taken to be
particles concentrated at a point. This leads to the idea of a *rate of
change of position*, i.e., if in a short time interval δt, a particle has
moved through a length δs, then we define its rate of change of
position averaged over the time interval δt as

$$\mathbf{v}_{av} = \frac{\delta \mathbf{s}}{\delta t} \tag{1.1}$$

and call it the *average velocity*. We define the true *velocity* at any time
as the limit of this expression as $\delta t \to 0$, thus,

$$\mathbf{v} = \lim_{\delta t \to 0} \frac{\delta \mathbf{s}}{\delta t}, \tag{1.2}$$

i.e., as the gradient of the tangent to a distance–time graph. Considerations
like these led Newton (1642–1727) to the invention of the infinitesimal
calculus, which was invented simultaneously and independently by
Leibnitz (1646–1716) in Germany. We shall use the notations developed
by both of them and write

$$\mathbf{v} = \frac{d\mathbf{s}}{dt} = \dot{\mathbf{s}}. \tag{1.3}$$

(The dot notation invariably refers to differentiation with respect to
time.)

Solution

1.   (A)   If this is not your answer, go
back to p. xvi now and
make sure that you understand
all there is in the section,
including the flow diagram
on progress tests and the
Greek alphabet.

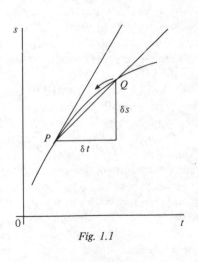

Fig. 1.1

Throughout the preceding discussion we have printed s and **v** in heavy type to indicate that they are *vectors*. A vector is a quantity that has both magnitude and direction (it is important not only to know how fast one is going, but where one is going), while a quantity like time interval, which has only magnitude, is called a *scalar*. Another scalar is *speed*, which is the magnitude of a velocity. The addition of vectors takes place through the *parallelogram law*. This is obvious for the displacement vector, since the position P of a point obtained through the displacement $\overrightarrow{OP}$ is clearly the same as that obtained through the successive displacements $\overrightarrow{OQ} + \overrightarrow{QP}$ or $\overrightarrow{OR} + \overrightarrow{RP}$. Note that such a vector does not have a particular line of action in this diagram so that $\overrightarrow{OQ} = \overrightarrow{RP}$. As velocities are obtained by dividing displacements by the scalar time interval, the parallelogram law holds for them too. Note also that the general vector addition,

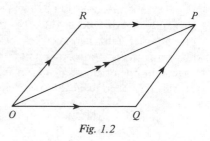

Fig. 1.2

$$\mathbf{A} + \mathbf{B} = \mathbf{C}, \tag{1.4}$$

can be most simply described by the vector triangle of addition. The magnitude of a vector **A** is denoted by A.

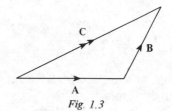

Fig. 1.3

In a similar way to velocity, we define *acceleration* as the rate of change of velocity, that is,

$$\mathbf{a} = \frac{d\mathbf{v}}{dt} = \dot{\mathbf{v}}$$

or

$$\mathbf{a} = \frac{d^2\mathbf{s}}{dt^2} = \ddot{\mathbf{s}}. \tag{1.5}$$

The units of velocity and acceleration are clearly m s^{-1} and m s^{-2} respectively, and their dimensions in terms of length L and time T are

$$[\mathbf{v}] = LT^{-1},$$

$$[\mathbf{a}] = LT^{-2}. \tag{1.6}$$

Conversely, if we know the velocity **v** of a point at all times t of a motion, then we can find the distance s covered in, say, the interval from time t_1 to time t_2, by dividing this interval into very small intervals δt. For each of these the distance covered is δs, where, from (1.3)

$$\delta\mathbf{s} = \mathbf{v}\,\delta t. \tag{1.7}$$

The total distance is then obtained by summing over all the small intervals and going to the limit,

$$\mathbf{s} = \lim_{\delta t \to 0} \sum_{t_1}^{t_2} \mathbf{v}\,\delta t = \int_{t_1}^{t_2} \mathbf{v}\,dt, \tag{1.8}$$

in the usual notation of the integral calculus. That the definite integral is the limit of a sum (the integral sign is an elongated S) is a property which we shall use frequently, and in this case it will be useful to distinguish the sum Σ (the sum symbol is the Greek letter sigma) from the integral symbol. For motion in one dimension, we can plot a velocity–time graph, and distance is then the area under the curve.

Another important vector to be defined is the *position vector* **r**, which gives the position of a point P relative to a fixed reference point O, generally the origin of a co-ordinate system. If in time δt, the point P moves to Q, then its position vector increases by δ**r**, so that

$$\overrightarrow{PQ} = \delta\mathbf{r}.$$

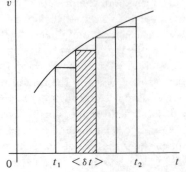

Fig. 1.4

Hence the velocity of P is given by

$$\mathbf{v} = \lim_{\delta t \to 0} \frac{\delta\mathbf{r}}{\delta t},$$

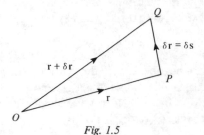

Fig. 1.5

and the infinitesimal displacement vector δ**s** is equal to the change in the position vector δ**r**.

In the rest of the book, we shall allow ourselves one simplification of notation. Although after going to the limit, ds/dt is one expression and not a fraction with a numerator and denominator, before going to the limit it is exactly that. Mathematicians are very particular to distinguish ds/dt from δs/δt, but we shall find that it does not cause us any confusion, and in fact is helpful to thought, to write the small quantity, say δs, as ds even before going to the limit. This will also make it possible for us to use quantities such as ds by themselves, as small quantities, which may or may not be infinitesimal.

Progress test

1. A velocity of 20 m s^{-1} is equal to about

   A   15 miles/hour
   B   30 miles/hour
   C   45 miles/hour
   D   60 miles/hour.

2. Which of the following is not an expression for acceleration?

   A $\dot{s}\dfrac{d\dot{s}}{ds}$

   B $\ddot{s}$

   C $\dot{s}\dfrac{d\dot{s}}{dt}$

$$\text{D} \quad \frac{d\dot{s}}{dt}$$

3. Distance is the area under the graph of

   A   velocity against time
   B   acceleration against velocity
   C   acceleration against time
   D   none of these.

1.4 Uniform motion in a straight line and in a circle

Two very simple kinds of motion are of special importance. The first of these occurs when the acceleration is constant. In that case the velocity changes uniformly and we have for the velocity $\mathbf{v}_2$ at time t_2 in terms of the velocity $\mathbf{v}_1$ at time t_1,

$$\mathbf{v}_2 = \mathbf{v}_1 + \mathbf{a}(t_2 - t_1). \tag{1.9}$$

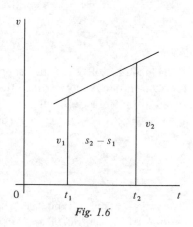

Fig. 1.6

Similarly, on considering the area under the graph,

$$\mathbf{s}_2 = \mathbf{s}_1 + \mathbf{v}_1(t_2 - t_1) + \tfrac{1}{2}\mathbf{a}(t_2 - t_1)^2. \tag{1.10}$$

On eliminating $t_2 - t_1$, we obtain

$$v_2^2 = v_1^2 + 2\mathbf{a} \cdot (\mathbf{s}_2 - \mathbf{s}_1), \tag{1.11}$$

where we have introduced the so-called *scalar product* of two vectors. For two vectors, $\mathbf{A}$ and $\mathbf{B}$, this is defined as

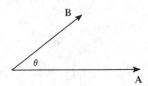

Fig. 1.7

$$\mathbf{A} \cdot \mathbf{B} = AB\cos\theta, \tag{1.12}$$

where θ is the angle between the vectors.* (The elimination of $t_2 - t_1$ is not quite so easy when vectors are involved. See worked example 1.1. Note that you are not allowed to divide by a vector.)

    Another important motion is that in which a point moves with *uniform speed in a circle*, so that the magnitude of the velocity is constant, but not its direction. Consider that a particle has moved with speed v from P to Q along the arc of a circle of radius r in time dt. Then the velocity changes from $\mathbf{v}_1$ to $\mathbf{v}_2$, where both vectors have the magnitude v. The vector triangle for the addition

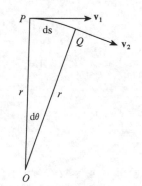

Fig. 1.8

$$\mathbf{v}_1 + d\mathbf{v} = \mathbf{v}_2 \tag{1.13}$$

is therefore isosceles, and in the limit $d\mathbf{v}$ becomes perpendicular to $\overrightarrow{AB}$, i.e., parallel to $\overrightarrow{PO}$. Hence the acceleration $\mathbf{a} = \dot{\mathbf{v}}$ acts along $\overrightarrow{PO}$. We

* Note that $\mathbf{A} \cdot \mathbf{A}$ is often written $\mathbf{A}^2$ and is equal to A^2.

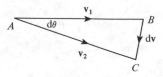

Fig. 1.9

obtain its magnitude by observing that the triangles OPQ and ABC are similar. Hence

$$\frac{\mathrm{d}s}{r} = \frac{\mathrm{d}v}{v}, \quad \text{that is,} \quad \frac{\mathrm{d}v}{\mathrm{d}s} = \frac{v}{r}.$$

Therefore

$$a = \frac{\mathrm{d}v}{\mathrm{d}t} = \frac{\mathrm{d}v}{\mathrm{d}s}\frac{\mathrm{d}s}{\mathrm{d}t} = \frac{v^2}{r}. \tag{1.14}$$

The final result is therefore that in uniform circular motion with speed v, the acceleration, which is called the *centripetal acceleration*, is directed towards the centre of the circle and is of magnitude v^2/r.

The *angular velocity*, or more accurately, the angular speed, is defined as

$$\omega = \frac{v}{r}. \tag{1.15}$$

It is measured in rad s^{-1} and has the dimension T^{-1}. In terms of it, the magnitude of the centripetal acceleration is

$$a = r\omega^2. \tag{1.16}$$

Worked example 1.1. To eliminate $t_2 - t_1$ from eqs. (1.9) and (1.10).
    We form the scalar product of $\mathbf{a}$ with (1.10):

$$\mathbf{a}\cdot(\mathbf{s}_2 - \mathbf{s}_1) = \mathbf{a}\cdot\mathbf{v}_1(t_2 - t_1) + \tfrac{1}{2}a^2(t_2 - t_1)^2.$$

(Note that $\mathbf{a}\cdot\mathbf{a} = a^2$, from the definition of the scalar product.)
Now from (1.9),

$$\mathbf{a}(t_2 - t_1) = \mathbf{v}_2 - \mathbf{v}_1 \quad \text{and} \quad a^2(t_2 - t_1)^2 = (\mathbf{v}_2 - \mathbf{v}_1)\cdot(\mathbf{v}_2 - \mathbf{v}_1).$$

Hence

$$\mathbf{a}\cdot(\mathbf{s}_2 - \mathbf{s}_1) = (\mathbf{v}_2 - \mathbf{v}_1)\cdot\mathbf{v}_1 + \tfrac{1}{2}(\mathbf{v}_2 - \mathbf{v}_1)\cdot(\mathbf{v}_2 - \mathbf{v}_1)$$
$$= \mathbf{v}_2\cdot\mathbf{v}_1 - v_1^2 + \tfrac{1}{2}v_2^2 + \mathbf{v}_2\cdot\mathbf{v}_1 + \tfrac{1}{2}v_1^2$$

or

$$2\mathbf{a}\cdot(\mathbf{s}_2 - \mathbf{s}_1) = v_2^2 - v_1^2$$

which is (1.11).

Solutions

1. (C) 45 miles per hour = 66 ft s^{-1} ≈ 20 m s^{-1} [10 ft ≈ 3 m]

2. (C) $\mathrm{d}s/\mathrm{d}t = \dot{s}$, which is acceleration.

3. (A) As $v = \mathrm{d}s/\mathrm{d}t$, $s = \int v\,\mathrm{d}t$. Hence $\dot{s}\ \mathrm{d}s/\mathrm{d}t$ is not.

Progress test

1. For motion under no acceleration, which one of the following is not constant?

   A   position
   B   speed
   C   velocity
   D   retardation.

2. The magnitude of the acceleration in uniform circular motion is

   A v/r
   B v^2/r
   C $r^2\omega$
   D $r\omega$.

3. Angular velocity is measured in

   A   radians
   B   degrees
   C   rad s^{-1}
   D   s^{-1}.

1.5   Relative motion

A matter that often causes difficulty is the motion of one point P_1 relative to another, P_2, which in turn is in motion. If the velocity of P_1 is v_1 and that of P_2 is v_2, then the velocity of P_1 relative to P_2 is denoted by v_{12}, where the letter is defined as

$$\mathbf{v}_{12} = \mathbf{v}_1 - \mathbf{v}_2. \qquad (1.17)$$

Note that the velocity of P_2 relative to P_1 is given by

$$\mathbf{v}_{21} = -\mathbf{v}_{12}. \qquad (1.18)$$

In order to avoid confusion it is best to avoid the phrase 'relative velocity' and always to speak of the velocity of one point relative to another. Similarly the position vector of P_1 relative to P_2 is denoted by $\mathbf{r}_{12}$, where

$$\mathbf{r}_{12} = \mathbf{r}_1 - \mathbf{r}_2$$

and $\mathbf{r}_1$ and $\mathbf{r}_2$ are the position vectors of P_1 and P_2 relative to an origin O.

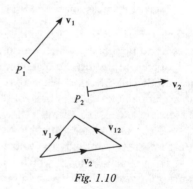

Fig. 1.10

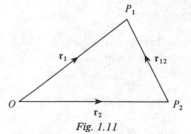

Fig. 1.11

Progress test

1.   If two particles P_1 and P_2 move with velocities $\mathbf{v}_1$ and $\mathbf{v}_2$, then the velocity of P_1 relative to P_2 is

A $\mathbf{v}_1 + \mathbf{v}_2$
B $\mathbf{v}_1 - \mathbf{v}_2$
C $\mathbf{v}_2 - \mathbf{v}_1$
D   none of these.

1.6   Dynamics, the study of what causes motion

We now introduce the concept of force. We take this to be a primary concept like length and time. Sommerfeld (1868–1951) writes in his *Mechanics,* Academic Press, 1964:

'We believe that we possess through our muscles an immediate impression of the concept of force, although it may only be qualitative. Beyond this, we have everywhere on the earth gravity, which we can use as a measuring standard, in order to measure force quantitatively. It is only necessary to balance a given force against standard weights.'

Other writers believe that *Newton's first law*,

'Every body remains in its state of rest or uniform motion in a straight line except in so far as it is compelled to change that state by forces impressed on it',

defines force, but it is rather a way of defining 'absence of force' and gives us no method by which we can measure force.

From our intuitive understanding of force, we see that it has both magnitude and direction. We shall therefore treat it as a vector, although we are not able to show at present that forces can be added in the way vectors are added, i.e., through the vector triangle of addition, Fig. 1.3.

The 'motion', mentioned in the first law, is defined by Newton as being measured jointly by the velocity and the *mass* of the body. As we have not yet defined mass, which most emphatically is not as basic a concept immediately available to our senses as is force, we instead define the *motion*–in modern terms *momentum*–through *Newton's second law*,

'The rate of change of momentum of a body is proportional to the force impressed on it and is effective in the direction in which that force is impressed.'

As the rate of change of momentum is proportional to a force, which is a vector, say $\mathbf{F}$, it too is a vector, and so must be momentum itself, which we denote by $\mathbf{p}$. The unit of momentum is defined in terms of the unit

Solutions

1.   (A)   If there is no acceleration, then there is no retardation (which is negative acceleration), and velocity and speed are constant.

2.   (B)   The two expressions for the centripetal acceleration are v^2/r and $r\omega^2$.

3.   (C)   Angles are measured in radians.

of force (so far unspecified) by making the proportionality into an equality,

$$\frac{d\mathbf{p}}{dt} = \mathbf{F}. \tag{1.19}$$

This both defines momentum and states its fundamental properties.

Newton's third law states that 'An action is always opposed by an equal and opposite reaction; or, the mutual actions of two bodies are always equal and act in opposite directions.'

Using this law, we can apply different forces to the same body and measure both the force and the resulting acceleration each time. They are found to be proportional, and we define the proportionality constant m in the equation

$$\mathbf{F} = m\mathbf{a} \tag{1.20}$$

as the *mass* of the body. Comparing this with (1.19), we then have

$$\mathbf{p} = m\mathbf{v}. \tag{1.21}$$

It may be noted that in going from kinematics to dynamics, we introduced one new primary concept, force, and were then able to define secondary concepts, such as momentum and mass, in terms of it. In strict logic, we could instead have started with either momentum or mass and then defined the other and force in terms of it. To that extent it may be considered a matter of taste, which of the three is taken to be primary, but then, much of physics is a matter of taste.

If forces $\mathbf{F}_1, \mathbf{F}_2$ act between two bodies in isolation, then the third law states that

$$\mathbf{F}_1 + \mathbf{F}_2 = 0, \tag{1.22}$$

or

$$\frac{d}{dt}(\mathbf{p}_1 + \mathbf{p}_2) = 0,$$

which leads to

$$\mathbf{p}_1 + \mathbf{p}_2 = \text{constant}, \tag{1.23}$$

or

$$m_1\mathbf{v}_1 + m_2\mathbf{v}_2 = m_1\mathbf{v}_1' + m_2\mathbf{v}_2', \tag{1.23'}$$

where $\mathbf{v}_1, \mathbf{v}_2$ and $\mathbf{v}_1', \mathbf{v}_2'$ are the velocities of two bodies of masses m_1 and m_2, at different times. This is the *law of conservation of momentum* for an isolated system of two bodies. It has been found that this law remains valid, even when Newton's third law breaks down, e.g., for the interaction between an electric charge and a current. Thus the law of conservation of momentum is more fundamental than the third law.

Historically, Newton's three laws form the introduction to his *Principia*, the great work which first systematized the motion of bodies. Even today, after nearly 300 years, they still are the cornerstone of our study of dynamics.

Worked example 1.2. An electron, velocity v_0, enters the gap between the deflector plates of a cathode ray oscilloscope, as shown. If the electric field between the plates is E, find the deflection y on the screen.

(i) Let the electron have mass m and charge $-e$, where e is positive, and assume that the field between the deflector plates is uniform. Then the force on the electron is $F = eE$ towards the positive plate, and therefore at right angles to the initial direction of motion. Hence there is no acceleration in the horizontal direction, and the time taken to traverse the distance between the plates is

$$t_1 = \frac{d}{v_0}.$$

During this time the vertical distance moved through is

$$y_1 = \frac{1}{2}\frac{eE}{m}t_1^2 = \frac{eEd^2}{2mv_0^2},$$

and the vertical component of the velocity at time t_1 is

$$v_1 = \frac{eE}{m}t_1 = \frac{eEd}{mv_0}.$$

(ii) The electron now enters a force-free region and reaches the screen after a further time

$$t_2 = \frac{l}{v_0}.$$

The vertical distance travelled during that time is

$$y_2 = v_1 t_2 = \frac{eEdl}{mv_0^2}.$$

The total deflection is therefore

$$y = \frac{eEdl}{mv_0^2}\left(1 + \frac{d}{2l}\right).$$

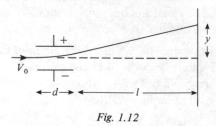

Fig. 1.12

Solution

1. (B) Are you clear why it is not $v_2 - v_1$? Construct a vector diagram.

Worked example 1.3. A particle with electric charge q and mass m, moving with velocity **v**, *enters a constant magnetic field, flux density* **B**, *at right angles to its direction of motion. Describe the subsequent motion.*

The force on the particle has the constant magnitude

$$F = qvB$$

and is at right angles to both **v** and **B** as shown. (We take q to be positive.) Hence it is always perpendicular to the direction of motion, so that the particle describes a circle in a plane perpendicular to **B**. The radius of the circle is obtained from the fact that F must give the centripetal acceleration (1.14), that is,

$$\frac{mv^2}{r} = F = qvB.$$

$$\therefore \quad r = \frac{mv}{qB}.$$

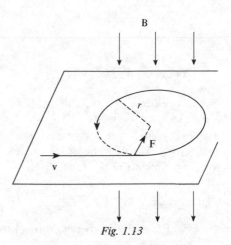

Fig. 1.13

Thus, for a given q and B, the radius is a measure of the momentum of the particle. The time for a complete revolution is

$$T = \frac{2\pi r}{v} = \frac{2\pi m}{qB}.$$

This is independent of the speed of the particle, a fact which is made use of in the construction of the cyclotron.

Worked example 1.4. A rocket is ejecting matter with constant relative velocity v_r. If it is to accelerate from velocity v_0 to velocity v_f, show that its initial mass m_0 is given by

$$m_0 = m_f e^{(v_f - v_0)/v_r}$$

where m_f is its final mass.

Let its mass and velocity at any instant be m and v, and let it eject a mass Δm in the next instant. Then the conservation of momentum equation gives

$$mv = (m - \Delta m)(v + dv) + \Delta m(v - v_r)$$

or

$$0 = m\,dv - v_r\,\Delta m.$$

Before integrating, we note that the change in mass of the rocket is $dm = -\Delta m$, since the rocket is losing mass. Hence

$$\frac{dm}{m} = -\frac{dv}{v_r}.$$

Integrating, we obtain

$$\ln \frac{m_0}{m_f} = \frac{v_f - v_0}{v_r},$$

or

$$m_0 = m_f e^{(v_f - v_0)/v_r}.$$

Exhaust velocities are typically of the order 3 km s^{-1}. Hence even to obtain a cruising velocity of 30 km s^{-1}, which would enable us to reach the outer planets in a few years, it is necessary for the weight of the fuel carried initially to be of the order of $e^{10} \simeq 20\ 000$ times the weight of the rest of the rocket!

Progress test

1. Which of the following is not a form of Newton's second law?

    A $\mathbf{F} = \dfrac{d\mathbf{p}}{dt}$

    B $\mathbf{F} = m\mathbf{f}$

    C $\mathbf{F} = m\dfrac{d\mathbf{v}}{dt}$

    D $\mathbf{F} = m\mathbf{v}.$

2. The law of conservation of momentum is valid for

    A   one body under an external force
    B   two bodies under an internal force between them
    C   two bodies under external forces
    D   two bodies under both internal and external forces.

1.7   The equation of motion

Let us return to eq. (1.20) and treat it in one dimension. In terms of the position co-ordinate s of a particle we can write it

$$m \frac{d^2 s}{dt^2} = F(s), \tag{1.24}$$

where the force F is a function of position, i.e., it varies from point to point. This differential equation giving the position $s(t)$ as a function of the time t of a particle at s moving under a force $F(s)$. Since it involves a second derivative, we have to integrate twice to obtain s. This introduces two constants of integration, generally determined by the position $s(0)$ and velocity $\dot{s}(0)$ at time $t = 0$. If we know these, then, as we shall see, we can find s at any subsequent time. For a small time interval, say τ, we have

$$\dot{s}(0) = \left[\frac{ds}{dt}\right]_{t=0} \simeq \frac{s(\tau) - s(0)}{\tau},$$

so that

$$s(\tau) \simeq s(0) + \tau\dot{s}(0). \qquad (1.25)$$

Similarly, starting with $\dot{s}(0)$,

$$\dot{s}(\tau) \simeq \dot{s}(0) + \tau\ddot{s}(0), \qquad (1.26)$$

where, from (1.24)

$$\ddot{s}(0) = \frac{F[s(0)]}{m}. \qquad (1.27)$$

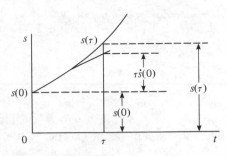

Fig. 1.14

In this way we can proceed step by small step and so find s at any time t. In principle, this is the method of integration of a differential equation employed by a computer, although in practice more accurate formulae than (1.25) and (1.26) are used. Since (1.25) and (1.26) are only approximations, which improve in quality as τ gets smaller, it is essential in this procedure to keep the time step τ as small as possible. On the other hand, the smaller τ, the longer the calculation to obtain $s(t)$ for a given t.

Note that the *equation of motion* (1.24) is very general. It covers all motions under the given force $F(s)$, different motions simply having different *initial conditions* $s(0)$ and $\dot{s}(0)$. Such generality is typical of differential equations.

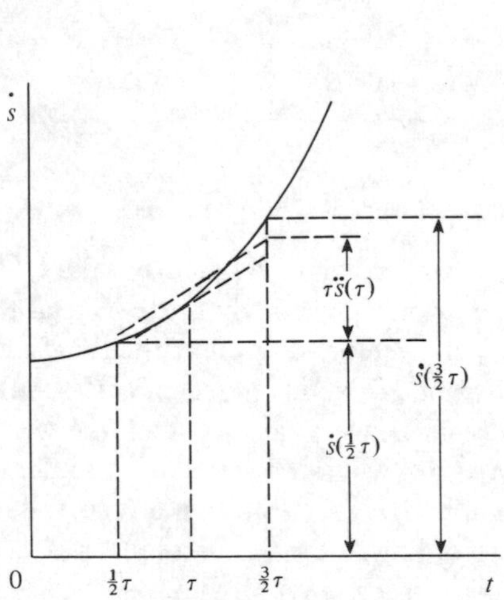

Worked example 1.5. Before giving an example of how to solve an equation of motion numerically, we show how the accuracy of (1.25) can be improved.

This is actually a Taylor series, which to the next term yields

$$s(\tau) \simeq s(0) + \tau\dot{s}(0) + \tfrac{1}{2}\tau^2\ddot{s}(0)$$

$$= s(0) + \tau[\dot{s}(0) + \tfrac{1}{2}\tau\ddot{s}(0)]$$

$$\therefore \quad s(\tau) \simeq s(0) + \tau\dot{s}(\tfrac{1}{2}\tau) \qquad (1.25')$$

using (1.26). Similarly, starting with $\dot{s}(\tfrac{1}{2}\tau)$

$$\dot{s}(\tfrac{3}{2}\tau) \simeq \dot{s}(\tfrac{1}{2}\tau) + \tau\ddot{s}(\tau). \qquad (1.26')$$

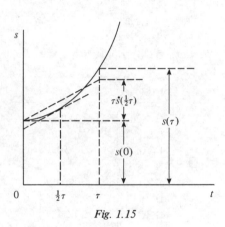

Fig. 1.15

Fig. 1.16

Thus, to do the calculation, we proceed in steps of τ, but calculate $\dot{s}$ at values halfway between those of s. The starting value $\dot{s}(\tfrac{1}{2}\tau)$ has to be obtained from the equation

$$\dot{s}(\tfrac{1}{2}\tau) \simeq s(0) + \tfrac{1}{2}\tau\ddot{s}(0).$$

Let us now solve the equation of motion for a particle of unit mass moving under an attractive force equal to ten times the square of the distance from the origin (in appropriate units), with initial conditions $\dot{s}(0) = 1$, $\dot{s}(0) = 0$. The equation, which incidentally cannot be integrated algebraically, is

$$\ddot{s} = -10s^2.$$

We put $\tau = 0\cdot1$ s, and set out the calculation in tabular form.

t	s	$\dot{s}$	$\ddot{s}$
0	1·000	0·000	− 10·000
0·05		− 0·500	
0·10	0·950		− 9·025
0·15		− 1·403	
0·20	0·810		− 6·561
0·25		− 2·059	
0·30	0·604		− 3·648
0·35		− 2·424	
0·40	0·362		− 1·310
0·45		− 2·555	
0·50	0·106		− 0·112
0·55		− 2·566	
0·60	− 0·151		+ 0·228
0·65		− 2·543	
0·70	− 0·405		+ 1·640 etc.

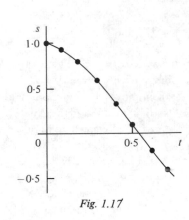

Fig. 1.17

1.8 Superposition of forces

It is an experimental fact that the effect of two forces acting at the same point on a given body is the same as the effect of the one force obtained by the vector addition of the two forces. In other words, if two forces $\mathbf{F}_1$ and $\mathbf{F}_2$, acting independently on the body, produce respective accelerations $\mathbf{a}_1$ and $\mathbf{a}_2$, then the force which produces an acceleration $\mathbf{a}_1 + \mathbf{a}_2$ (where we are allowed to use the vector addition law, since we know that acceleration is a vector) is obtained by adding $\mathbf{F}_1$ and $\mathbf{F}_2$ according to the vector addition law. This completes our demonstration that force is a vector and it should be noted that the vector nature of force cannot be deduced from Newton's second law, which only deals

Solutions

1. (D) mv is the momentum, not the rate of change of momentum.

2. (B) Only in this case is there no net resultant force, since the internal forces on the two bodies are equal and opposite.

with one force, or from the third law, which only deals with forces in direct opposition. The first person to realize that the vector nature of force required a postulate separate from Newton's laws was Mach (1838–1916).

1.9 Units and dimensions

The unit of mass is the kilogram (kg), defined through the mass of a standard lump of platinum.

The unit of force is the newton (N). It is that force which gives a mass of 1 kg an acceleration of 1 m s^{-2}.

The *weight W* of a body is the force with which the earth attracts it. It is found to be proportional to the mass, and we put

$$W = mg. \tag{1.28}$$

Comparing this with (1.20), we see that g is the acceleration which a body experiences as a result of its attraction by the earth. Clearly this quantity is different in different locations. In London,

$$g = 9{\cdot}81 \text{ m s}^{-2}.$$

In many rough calculations, we may take $g \simeq 10 \text{ m s}^{-2}$.

The proportionality of mass and weight has made it possible to compare masses by comparing corresponding weights. In fact, the chemists' law of conservation of matter was verified by measuring weights, not masses. We shall return to the problem of the direct measurement of mass and of the conservation of mass in the next chapter.

We have already stated that in going from kinematics to dynamics we had to add one new concept, and that we chose force for this. Correspondingly, we have to add one new basic unit, and it is conventional for this to be mass, M. In terms of this, the dimensions of force and momentum are

$$[F] = MLT^{-2}, \qquad [p] = MLT^{-1}. \tag{1.29}$$

An important use of the concept of dimensions concerns the dimensionality of different terms in a physical equation. For two physical quantities to be additive, it must be possible to measure them in the same units, and in mechanics this means that they must have the same dimensions in three basic units—mass, length and time. (When dealing with phenomena in electromagnetism and thermodynamics, two more basic units—for electric current and for temperature—have to be added.) It then follows that in any physical equation, the different terms must each have the same dimensions, i.e., the dimensional nature of each term in an equation in mechanics can be expressed in the form

$$M^r L^s T^t,$$

where r, s and t are numbers which are the same for all the terms in the equation. We shall use these considerations when dealing with energy in the next section and in section 2.11, and with viscosity in section 7.5.

Worked example 1.6. If the unit of force in the CGS system is the dyne, show that

$$1 \text{ dyn} = 10^{-5} \text{ N}.$$

The dimensions of force are

$$[F] = MLT^{-2}.$$

Hence

$$1 \text{ dyn} = \frac{1 \text{ g} \times 1 \text{ cm}}{(1 \text{ s})^2}$$

$$= \frac{10^{-3} \text{ kg} \times 10^{-2} \text{cm}}{(1 \text{ s})^2}$$

$$= 10^{-5} \text{ N}.$$

Progress test

1. A force of 1 newton is equivalent to the weight of a mass of about

    A   0·01 kg
    B   0·1 kg
    C   1·0 kg
    D   10   kg

2. The unit of pressure is 1 N m^{-2} in the MKS system and 1 dyn cm^{-2} in the CGS system. If 1 dyn = 10^{-5} N, then 1 N m^{-2} is

    A   0·1 dyn cm^{-2}
    B   1 dyn cm^{-2}
    C   10 dyn cm^{-2}
    D   100 dyn cm^{-2}.

1.10 Energy

If we multiply (1.11), which can be written

$$\mathbf{a} \cdot \mathbf{s}_2 - \mathbf{a} \cdot \mathbf{s}_1 = \tfrac{1}{2}v_2^2 - \tfrac{1}{2}v_1^2 \tag{1.11}$$

by m and use Newton's second law,

$$F = ma, \tag{1.20}$$

then we have, for constant acceleration,

$$\mathbf{F} \cdot \mathbf{s}_2 - \mathbf{F} \cdot \mathbf{s}_1 = \tfrac{1}{2}mv_2^2 - \tfrac{1}{2}mv_1^2. \tag{1.30}$$

The right-hand side of this equation depends only on the mass and the velocity of the body at each of two different times. Thus, the quantity

$$T = \tfrac{1}{2}\,\text{mass} \times (\text{velocity})^2 \tag{1.31}$$

is in some way a measure of the 'motion' of the body, which is different from its momentum. It is called the *kinetic energy*. The left-hand side depends only on the force and on the point at which the force is applied at the two different times. It does not depend on the motion—on the contrary, it is what causes the motion and hence changes the kinetic energy. The quantity

$$W = \overrightarrow{\text{force}} \cdot \overrightarrow{\text{distance}} \tag{1.32}$$

is called *work*. Thus

$$\text{work done on the body} = \text{increase in kinetic energy.} \tag{1.33}$$

It is easily verified that the dimensions of work, as defined in (1.32), and of kinetic energy, as defined in (1.31), are the same, as of course they must be. For that reason, they must also be measured in the same units. The unit of work and energy is the *joule*.

We can re-write (1.30) as a conservation law, if we define *potential energy V* through

$$V = -W \tag{1.34}$$

Then (1.30) can be written

$$T_1 + V_1 = T_2 + V_2$$

or

$$T + V = \text{constant.} \tag{1.35}$$

It follows from (1.28), that for a distance h above the surface of the earth, which is small enough for the weight of a body to be taken to be constant, the gravitational potential energy is

$$V = mgh. \tag{1.36}$$

So far we have derived (1.35) only for uniformly accelerated motion. Now it is always possible to write .

$$\frac{d^2 s}{dt^2} \quad \text{as} \quad \dot{s}\frac{d\dot{s}}{ds},$$

and hence for any motion, we can re-write (1.24) as

$$m\dot{s}\cdot\frac{d\dot{s}}{ds} = F(s). \qquad (1.37)$$

We now integrate this from an initial position s_0 at time $t = 0$ to an arbitrary position s at time t. To avoid confusion, we use s' for the variable of integration. Then

$$\tfrac{1}{2}m\dot{s}^2 - \tfrac{1}{2}m\dot{s}_0^2 = \int_{s_0}^{s} F(s')\,ds'. \qquad (1.38)$$

We next introduce a new quantity $V(s)$ through the equation

$$F(s) = -\frac{dV}{ds}. \qquad (1.39)$$

Then

$$\int_{s_0}^{s} F(s')\,ds' = -V(s) + V(s_0). \qquad (1.40)$$

$$\therefore \quad \tfrac{1}{2}m\dot{s}^2 - \tfrac{1}{2}m\dot{s}_0^2 + V(s) - V(s_0) = 0, \qquad (1.41)$$

that is,

$$T(s) + V(s) = T(s_0) + V(s_0). \qquad (1.42)$$

Thus, the quantity $T(s) + V(s)$ does not depend on s and so is constant everywhere. Comparing with (1.35), we see that V is again the potential energy and that for a constant force F, the definition (1.39) reduces to that of (1.34), except that in our second proof we have confined ourselves to one dimension. The generalization of (1.39) to three dimensions, when F is not constant, will be dealt with in chapter 3.

   A force F which can be represented as the derivative of another function of position – $V(s)$, is called a *conservative force*. In one dimension, the condition for this to be possible is simply that F is a function of the position s, but in three dimensions we shall see that the condition is more complicated. For motion under conservative forces, the total energy–kinetic plus potential–is conserved. The function $V(s)$ contains an arbitrary constant of integration and this is usually chosen so that $V(\infty) = 0$. An exception is the gravitational potential energy (1.36) near the surface of the earth, when this surface is usually taken as the zero of potential energy.

   The idea that there is a quantity in nature, called energy, which is conserved has been considerably extended, so that we now have heat energy, electrical energy, nuclear energy and many other forms of energy. Physicists have a great love for conservation laws and give up almost anything rather than them. Before, however, adopting too reverent an attitude, you should read 'Dialogue on Statements and Mis-statements about Energy' by H. L. Armstrong, in *American Journal of Physics*, **33**, 1074 (1965).

Could we derive a corresponding generalized conservation law for momentum, remembering that (1.23) was valid only for an isolated system? If instead of multiplying (1.11) by $\frac{1}{2}m$, we multiply (1.9) by m and then use (1.20), we have

$$\mathbf{F}t_2 - \mathbf{F}t_1 = m\mathbf{v}_2 - m\mathbf{v}_1. \tag{1.43}$$

Clearly, we could now define a *potential momentum*, $-\mathbf{F}t$, and in terms of this establish a more general law for conservation of momentum for bodies under forces. It would appear that this is not a useful concept and so it has never been pursued. Instead, the product of force and time has, for infinitesimally short times and large forces been defined as the *impulse* **I**. Hence we can write (1.43) as

$$\mathbf{I} = m\,\mathbf{dv} = \mathbf{dp},$$

i.e., the impulse equals the change of momentum. This concept is useful, when the change of momentum is due, say, to a sharp blow, as when, for instance, a ball is struck by a bat.

Worked example 1.7. A body of mass m, at rest on a vertical spring of spring constant K, compresses the spring a distance b from its relaxed position. From what height must it be dropped to compress the spring a distance 3b?

From Hooke's law, the force to compress the spring a distance x is

$$F = Kx,$$

hence

$$mg = Kb.$$

The work done in compressing the spring a distance x is

$$W = \int_0^x F\,dx' = \int_0^x Kx'\,dx' = \tfrac{1}{2}Kx^2 = \frac{mg}{2b}x^2.$$

This is the potential energy stored in the spring, when it is compressed to this distance.

If the mass m is dropped from a height y above the relaxed position, then by the time it has compressed the spring a distance $3b$, it has lost potential energy $mg(y + 3b)$. As the spring and mass are then at rest, there is no kinetic energy, and the loss of gravitational potential energy must equal the gain in elastic potential energy in the spring. Hence

$$mg(y + 3b) = \frac{mg}{2b}(3b)^2$$

$$\therefore \qquad y = \tfrac{3}{2}b.$$

Progress test

1. Potential energy gained by a body acted on by a force is equal to

    A the work done on the body by the force
    B the work done by the body on the force
    C minus the work done by the body on the force
    D none of these.

2. For motion of a body under a conservative force

    A momentum is conserved, but not energy
    B energy is conserved, but not momentum
    C both energy and momentum are conserved
    D neither energy nor momentum are conserved.

3. The general expression for work done by a force **F** in moving its point of application through a distance s in time t is

    A $\mathbf{F} \cdot \mathbf{s}$
    B Fs
    C Ft
    D Ft.

4. Work of the order of 1 joule is done by a man lifting himself through

    A 1 mm
    B 1 cm
    C 10 cm
    D 1 m.

1.11 Energy diagrams

We can write (1.41) as

$$T(s) + V(s) = E,$$

where E is the constant *total energy* of the system. If we now plot $V(s)$ against s for a given E, then motion is possible only for those values of s for which $T(s) > 0$, since the kinetic energy is always positive. There are two specially simple kinds of force:

(a) force $F(s)$ everywhere attractive towards a given point which we take as origin, so that $F(s) > 0$ for $s < 0$ and $F(s) < 0$ for $s > 0$;

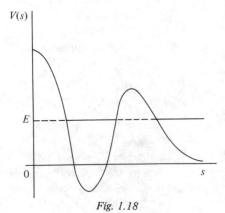

Fig. 1.18

(b)   Force $F(s)$ everywhere repulsive from a given point which we take
       as origin, so that $F(s) < 0$ for $s < 0$ and $F(s) > 0$ for $s > 0$.

Then for (a) the gradient of $V(s)$ is negative for $s < 0$ and positive for
$s > 0$, and conversely for (b). If we assume $V(\infty) = 0$, it then follows
that, for (a), the potential energy must be negative everywhere with a
minimum at $s = 0$, while conversely for (b) it must be positive everywhere with a maximum at $s = 0$. For that reason the two cases are
known as the potential well and potential barrier respectively. The
energy diagrams of Figs. 1.19, 1.20, show the motion of particles of
different total energies in the two cases. We define a *free state* as one in
which a particle is capable of reaching an infinite distance from the
force centre, a *bound state* as one in which it is not.

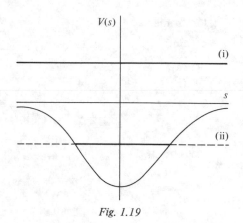

Fig. 1.19

(a)   *Well*
(i)   If a particle starts at infinity, it is first accelerated and then
       retarded by the well, ending up at infinity on the other side with
       initial speed (free state).
(ii)   Particle always remains at a finite distance from O and oscillates
       between two extreme positions (bound state).

(b)   *Barrier*
(i)   Particle coming from infinity is first retarded and then
       accelerated by the barrier, ending up at infinity on the other side
       with initial speed (free state)
(ii)   Particle coming from infinity is repelled by barrier, ending up at
       infinity on the same side with initial speed (free state).

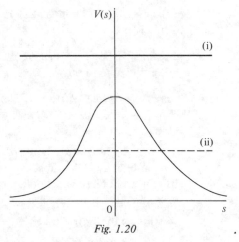

Fig. 1.20

    Note that, in all cases, the potential and kinetic energies of the
particle vary from point to point, but the total energy is constant.
    Applications of the above considerations occur frequently in
nuclear physics. The so-called *shell model* requires an understanding of
potential wells, while the *tunnel effect* relates to potential barriers.
[Jackson, chapter 7.]

Progress test

1.   In a potential barrier, the force is

      A   everywhere attractive towards a given point
      B   everywhere repulsive from a given point
      C   attractive from one side, but repulsive from the other
      D   none of these.

2.   If a particle coming from infinity meets a potential well, it is

      A   first retarded and then accelerated
      B   retarded throughout
      C   reflected
      D   first accelerated and then retarded.

1.12 Problems

1.1. What are the properties of two vectors **A** and **B** such that

(a) $\mathbf{A} + \mathbf{B} = \mathbf{C}$ and $A + B = C$
(b) $\mathbf{A} + \mathbf{B} = \mathbf{A} - \mathbf{B}$
(c) $\mathbf{A} + \mathbf{B} = \mathbf{C}$ and $A^2 + B^2 = C^2$
(d) $\mathbf{A} - \mathbf{B} = \mathbf{C}$ and $A + B = C$.

1.2. A rocket is fired vertically upwards with constant acceleration. After one minute it reaches a height of 36 km, and the motor is shut off.

(a) Draw a v–t diagram for the entire flight.
(b) Calculate the maximum height reached.
(c) Calculate the total time of flight.

[Take $g = 10 \text{ m s}^{-2}$ and neglect air resistance and variation of g with height.]

1.3. Find the direction and magnitude of the centripetal acceleration in London, due to the rotation of the earth about its axis.

1.4. A man can paddle a canoe at 5 km h^{-1} relative to the water and wishes to cross a river 100 m wide in which the speed of water is 3 km h^{-1}.

(a) In what direction should he paddle to reach the other side as quickly as possible?
(b) In what direction should he paddle if he wishes to land directly across the river?

How long will this trip take?

1.5. A passenger ship, travelling at 25 knots, observes a tanker 10 sea-miles away at an angle of 30° to its direction of motion. The tanker is travelling at 12·5 knots at 90° to the line of sight in such a way that the two ships approach each other. Show that the ships are on a collision course and calculate the time available to take evasive action. [1 knot = 1 seamile per hour.]

1.6 A body of mass 1 kg falls from a height of 100 m and reaches the ground with a speed of 40 m s^{-1}. What is the average air resistance? Discuss the meaning of 'average' in this connection.

1.7 A mass m is whirled at constant speed in a vertical circle at the end of a string of length l, the other end of which is kept fixed. What is the least angular velocity for the string not to become slack?

1.8. Show that if the motion of a particle of mass m is given by

$$s = s_0 \, e^{-kt} \sin \omega t,$$

Solutions

1. (B) The gradient is negative on the positive side and positive on the negative side of the maximum. Hence the force, which is the negative gradient, is always directed away from the maximum of the barrier.

2. (D) As in the previous example, consider the sign of the gradient for the two cases of the particle coming from $+$ or $-\infty$ respectively.

Solutions

1. (D) It is minus the work done on the body by the force, e.g., if a body falls, it loses P.E., but work is done on it by the force of gravity.

2. (B) If there is any resultant force acting, then momentum is not conserved.

3. (A) This is always correct, while (B) is correct only if F and s are in the same direction.

4. (A) A heavy man has a mass of 100 kg (No man's mass is either 10 or 1000 kg.). If he lifts himself through 1 mm, work done is $100 \times 10 \times 0.001 = 1$ J.

where s_0, k and ω are constant, then the equation of motion is

$$\ddot{s} + 2k\dot{s} + (k^2 + \omega^2)s = 0.$$

Interpret the different terms in this equation. Also sketch s as a function of t.

1.9. A body of mass m is resting on a table and fixed to two springs, spring constants K and $2K$, as shown. In the equilibrium position, the two springs are both relaxed. The mass is now moved a distance l to the left and released. Find (a) the distance to the right of the equilibrium position, when the mass next comes to rest and (b) the velocity of the mass as it passes through the equilibrium position. [You may take the table to be smooth.]

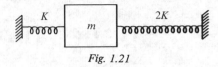

Fig. 1.21

1.10. A particle, mass m, starting from $-\infty$, approaches a force region the potential-energy function of which is

$$V = \frac{V_0}{x^2 + a^2},$$

where V_0 and a are constant.

(a) Derive an expression for the force on the particle.
(b) State in which region the force is attractive and in which repulsive.
(c) What is the least velocity at $-\infty$, which will allow the particle to reach $+\infty$?

Note added in proof: The application of Newton's laws to simple problems in circular motion and common errors that arise in the process are discussed by J. W. Warren, Circular motion, *Physics Education* **6**, 74 (1971).

2. Space, time and frames of reference

2.1. The nature of space and time

When we discussed the meaning of position of a point or time of an event in section 1.2, we concluded that we could define these only in relation to an arbitrary reference point and an arbitrary pointer reading on a clock. These arbitrary reference entities are jointly referred to as a *frame of reference* and in mathematical terms can be described through an orthogonal co-ordinate system *Oxyz* and a linear time scale *St*. Two questions immediately pose themselves:

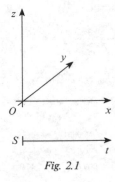

Fig. 2.1

(a)   Is our choice of frame of reference entirely arbitrary, or are some frames of references to be preferred to others?

(b)   Is there perhaps one frame of reference, which can be distinguished from all others and be described in some sense as absolute?

The weight of philosophical argument over the centuries inclined towards answering the second question in the affirmative, and Newton himself at the beginning of the *Principia* gives the following definitions:

> 'Absolute space, in its own nature, without regard to anything external, remains always similar and immovable.' 'Absolute, true and mathematical time', of itself and from its own nature, flows equally without regard to anything external.'

From these very definitions, it is clear that no experiment can be designed to observe the effects of absolute space and time, since nothing we can do will change them. They are therefore irrelevant to our description of nature, and Newton himself conceded that in experiments we use 'relative space' and 'relative time', as indicated in section 1.2. The fact that Newton added the rather puzzling adjectives 'true and mathematical' to his definition of absolute time may even indicate that he himself had qualms about the introduction of these absolute concepts.

2.2 Inertial frames of reference

Let us now turn to the other question. Although there is no evidence for the existence of absolute space and time, the existence of a common experience of the flow of time has until recently been essentially taken for granted. [See section 2.9.] For that reason, the term 'frame of reference' is often loosely used to refer merely to the space part, the common time part being taken for granted. Regarding the space part, we intuitively tend to refer external events to ourselves as a frame of reference, so that in a boat on a choppy sea we speak of the horizon going up or down, and when a car rounds a corner, we speak of a centrifugal force pushing us against the car door. We are aware that in some sense this centrifugal force is not a 'real' force, although if the door is not securely closed, the force can throw us out of the car and kill us! Our commonsense experience looks like being inadequate to cope with the problem.

The answer lies in Newton's first law. This involves a description of the motion of a body and, as we have seen, this can only be done in relation to a frame of reference. If the law turns out to be valid in relation to one particular frame, will it be equally valid in relation to all others? Clearly not, since for instance the centrifugal force observed by the driver of a car rounding a corner is not observed by an onlooker on the roadside. It was for this reason – that the force in some way depended on the motion of the observer – that we felt it was not a 'real' force. We therefore single out a particular frame of reference, for which Newton's first law is valid, and call it an *inertial frame of reference*. The law then reduces to the statement that such a frame exists, and that—because we cannot distinguish between rest and different uniform motions—all frames of reference in uniform motion relative to the first inertial frame are also inertial frames.

How to find the first inertial frame is not obvious. Experimentally, what is required is the verification of Newton's first law by observations over a period of time on a body which is not acted on by any forces. If, relative to a particular frame of reference, such a body moves with uniform velocity, then the frame is an inertial frame. But how can we be sure that a body is not acted on by any forces? Perhaps a rocket moving in free space far from all stars might fulfil this condition, and move uniformly relative to the stars. This cannot of course be verified, but it is plausible, and the system of fixed stars has indeed been used as an inertial frame, without leading so far to contradictions with experience. Because of the slowness of the rotation of the earth relative to the fixed stars, a terrestrial laboratory is a good approximation to an inertial frame and is frequently used as such. However, it is clear that the search for a definitely inertial frame of reference has not been entirely successful, and for many years the hope that there existed an inertial frame distinguished from all others kept the search for an absolute frame of reference alive.

Now that Newton's first law has led us to the concept of an inertial frame of reference, the status of his second and third law relative to an inertial frame must be considered. In fact, innumerable experiments in the laboratory—which we have stated is a good approximation to an inertial frame—have verified both laws to a high degree of accuracy. We therefore state that the second and third laws are valid for inertial frames of reference.

Progress test

1. A frame of reference can be described in terms of

   A   an orthogonal co-ordinate system $Oxyz$
   B   a time scale St
   C   an orthogonal co-ordinate system $Oxyz$ and a time scale St
   D   none of these.

2. Which one of the following does not require a frame of reference for its description?

   A   momentum
   B   kinetic energy
   C   mass
   D   position.

3. Which one of the following is not correct for motion relative to an inertial frame?

   A   Bodies under no forces are at rest
   B   The acceleration of a body is proportional to the force acting on it
   C   To every action there is an equal and opposite reaction
   D   Fixed stars move with uniform velocity.

2.3   Real and inertial forces

Let us now consider a spatial frame of reference which moves with accelerated motion relative to an inertial frame. In particular, consider $O'x'y'z'$ moving with acceleration $\mathbf{f}_0$ relative to the inertial frame $Oxyz$, and let observers O and O' situated on their respective origins of co-ordinates measure the acceleration of a particle of mass m at a point P relative to themselves as $\mathbf{f}$ and $\mathbf{f}'$ respectively. Then as O is an *inertial observer*, Newton's second law is valid for him and he concludes that there is a force $\mathbf{F}$ acting on the particle, given by

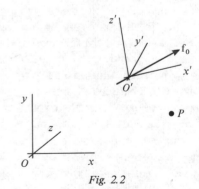

Fig. 2.2

$$\mathbf{F} = m\mathbf{a}. \tag{2.1}$$

Observer O' is not an inertial observer, but long habit has made him also associate a force with a measured acceleration. He therefore concludes that a force $\mathbf{F}'$ is acting on the particle, given by

$$\mathbf{F}' = m\mathbf{a}'. \tag{2.2}$$

But, it follows from the equation for relative velocity (1.17) that

$$\mathbf{a} = \mathbf{a_0} + \mathbf{a'}; \tag{2.3}$$

hence

$$\mathbf{F'} = \mathbf{F} - m\mathbf{a_0}. \tag{2.4}$$

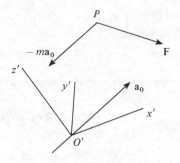

Fig. 2.3

According to O', the particle therefore experiences an additional force $-m\mathbf{a_0}$, which is not due to any external agent, but to the accelerated motion of O'. We call this an *inertial force*, because when O' experiences it on himself, it is due to his own inertia, i.e., his tendency to move with uniform velocity in an inertial frame, unless acted on by an external force. (From the point of view of the inertial roadside observer, people fall out of cars at corners, because they continue in a straight line, while the car does not. The force that turns the car is the frictional force between the tyres and the road, which is equal and opposite to the centrifugal force as experienced by the non-inertial observer in the car. More of this in chapter 6. The relevance of rotational motion to the problem of absolute space is discussed in Born, *Einstein's Theory of Relativity*, chapter 3, section 9.)

If the acceleration $\mathbf{a_0}$ is uniform, then the resulting inertial force $-m\mathbf{a_0}$ is clearly indistinguishable from a change of weight. If the true weight of a body, as measured by O, is $m\mathbf{g}$, then the weight as measured by O' is $m(\mathbf{g} - \mathbf{a_0})$. This effect is easily observable in a moving lift. For freely falling observers, $\mathbf{a_0} = \mathbf{g}$. We say that they are in a weightless condition.

At this point readers are strongly advised to see the PSSC film *Frames of Reference*. No amount of words can compare with this film in giving an understanding of frames of reference and of inertial forces.*

Progress test

1.   A man in a lift experiences an increase in weight if the velocity v and acceleration f of the lift are:

A $v \downarrow f \uparrow$
B $v \downarrow f \downarrow$
C $v \uparrow f = 0$
D $v \uparrow f \downarrow$.

* The film can be obtained from Sound Services Ltd., 269 Kingston Road, London, S.W.19.

2. An astronaut in an orbiting satellite experiences weightlessness because

    A   he is too far from the earth to feel its gravitational attraction

    B   the attraction from the earth is balanced by that from the moon

    C   the satellite is a non-inertial frame of reference

    D   he is falling freely.

3. Which one of the following is an inertial force?

    A   The force of gravity on a falling weight

    B   The force that throws a passenger forward when a car brakes suddenly

    C   The pull at one end of a string when a stone attached to the other end of it is whirled around.

    D   The force of the moon on the oceans that produces the tides.

2.4. Co-ordinate transformations

If two observers O and O' make observations of the same physical event, then there must be a relation between the measurements which they make. In particular, if O observes an event to occur at point (x,y,z) at time t, and O' observes the same event to occur at point (x',y',z') and at time t', then there must be a relation between (x,y,z,t) and (x',y',z',t'), each of which is adequate to describe the event. (Note that we use the word *event* to describe something that happens in both space and time.) Because we are considering one definite event observed by two observers, this relation must be 1–1, i.e., knowledge of one set of co-ordinates must enable us to predict the other uniquely, and conversely.

    Consider O' moving with uniform velocity $\mathbf{v}$ relative to O. It follows from (1.17) that the position co-ordinates $\mathbf{r}$ and $\mathbf{r}'$ of a point P relative to O and O' are related by

$$\mathbf{r}' = \mathbf{r} - \mathbf{v}t. \tag{2.5}$$

If the orthogonal axes $O'x'y'z'$ are taken parallel to $Oxyz$, then this can be written

$$x' = x - v_x t$$
$$y' = y - v_y t \tag{2.5'}$$
$$z' = z - v_z t$$

where (v_x, v_y, v_z) are the components of $\mathbf{v}$ along $Oxyz$. This is the required relation, which clearly satisfies the 1–1 condition. It is completed by the trivial relation

$$t' = t, \tag{2.6}$$

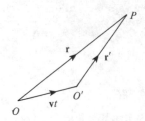

Fig. 2.4

based on the assumption that clocks are unaffected by motion. Equations (2.5) and (2.6) together are referred to as a *co-ordinate transformation*.

Since co-ordinate transformations are concerned with the recording of the same physical event by different observers, we expect there to be transformation equations also between physical observables, such as for instance force. We have already had such a relation in (2.4). In the present case, when **v** is constant, we have

$$\mathbf{F} = m\ddot{\mathbf{r}} = m\ddot{\mathbf{r}}' = \mathbf{F}', \tag{2.7}$$

i.e., the two observers measure the same force. Thus, Newton's laws are valid in the second frame, if they are valid in the first. Such frames are said to be *dynamically equivalent*, and an expression that is unchanged by the transformation, such as **F** in (2.7), is called an *invariant* of the transformation. It should be noted that two such frames need not necessarily be inertial, but that it follows from our definition of inertial frames that all inertial frames are dynamically equivalent. This is expressed in a principle, known as the *principle of Newtonian relativity*:

> 'In any given space region there exists an infinite set of inertial frames in which the laws of dynamics take on a specially simple form. It is impossible to distinguish between two such frames of reference by any dynamical experiment.'

In fact, this principle was in essence already known to Galileo who stated that it was impossible to detect the motion of a sailing boat drifting on a calm sea by any observation confined to the boat. Newton used it implicitly, when he based his treatment of dynamics on relative measurements of space and time. But the fact that he also introduced definitions for absolute space and time shows that he did not fully recognize its significance. This was not fully realized until the 19th century and the above formulation of the principle is a modern one. It has been named after Newton both out of deference to his great work and because it is implicit in it. For a similar reason the co-ordinate transformation (2.5) and (2.6) is called the *Galilean transformation*, although Galileo never used it explicitly.

Progress test

1. If an observer O' moves with a constant velocity **v** relative to another observer O, then the position vector $\mathbf{r}'$ at any time t, as measured by O', is given in terms of **r**, as measured by O, by

   A $\mathbf{r}' = \mathbf{r} + \mathbf{v}t$

   B $\mathbf{r}' = \mathbf{r} - \mathbf{v}t$

   C $\mathbf{r}' = \mathbf{r} + \mathbf{v}/t$

   D $\mathbf{r}' = \mathbf{r} - \mathbf{v}/t.$

2. If the two observers in the above question make corresponding observations of the same events, which one of the following is not correct?

A $\mathbf{F}' = \mathbf{F}$

B $m' = m$

C $\dot{\mathbf{r}}' = \dot{\mathbf{r}}$

D $\ddot{\mathbf{r}}' = \ddot{\mathbf{r}}$.

3. Which one of the following is not invariant under a general Galilean transformation?

A   The distance between two points
B   The magnitude of the force on a body
C   The component along the x-axis of the force on a body
D   The interval in time between two events.

2.5   The problem of the ether

Although the principle of Newtonian relativity has so far been tested only through dynamical experiments, it gives every appearance of being a universal principle of nature, so that a natural extension of it would be that it is impossible to distinguish between different inertial frames by *any experiment whatsoever*. This however was not the conslusion reached by the physicists of the 19th century who looked into other branches of physics and, in particular, optics in order to devise experiments which would distinguish between uniform motion and rest, and thereby define an absolute frame of reference.

The classical experiments on interference and diffraction of light of Thomas Young (1773–1829) and others had shown that light had the characteristics of a wave motion, and this led to the idea of an *ether*, the sole property of which was to act as a carrier of the light waves. Since the planets in their motion obeyed Newton's laws and thus were unaffected by this ether, the latter had to be exceedingly tenuous. However, by measuring the velocity of light from a source in different directions, we should be able to determine the velocity of the source through this ether and, since the ether was universal, this velocity would in fact be the absolute velocity of the source. (The same procedure, applied to sound, certainly does enable us to measure the velocity of a sound source relative to the surrounding air, since the velocity of the sound relative to the source and the velocity of the source relative to the air add according to the addition law of relative velocities (1.17) to give the velocity of the sound relative to the air.) This experiment was eventually performed by Michelson and Morley in 1887 (see problem 2.3), with the result that within the accuracy of the experiment the velocity of light was the same in all

Solutions

1. (A) An apparent increase in weight can only be due to an upward acceleration.

2. (D) The converse of the answer to question 1 is that a downward acceleration leads to an apparent loss of weight, so that a freely falling person appears to be weightless. Note that the astronaut 'falls' and yet gets no nearer to the earth.

3. (B) The passenger refers himself to the non-inertial frame of the car. In the inertial frame of the road he continues in uniform motion until he hits the decelerating windscreen.

directions. It is one of the classic experiments in physics and historically
was the experiment which first shook the general belief in the existence
of an ether. It has been repeated on a number of occasions since, mostly
with the same result, although there have been arguments in some
instances that a motion through the ether was detected. These have
never been substantiated. [The history of the experiment is reviewed in
R. S. Shankland, 'The Michelson–Morley experiment', *SA* 321, Novem-
ber 1964, p. 107.] A number of astronomical considerations showed
that this could not be due to the fact that the ether was dragged along
with the earth, and hence stationary relative to it. Further, the suggestion
that the velocity measured was that relative to the source was finally
disproved by an experiment on the decay of neutral pi-mesons in flight.
These particles, which arise in nuclear reactions at high energy, emit
electromagnetic radiation in the form of γ-rays, and it was observed
that pi-mesons travelling at speeds of 99·9 per cent of the speed of light
emitted γ-rays in the forward direction with speed equal to that of the
speed of light as measured using stationary sources in the laboratory.

While the negative result of the Michelson–Morley experiment
was at the time a cause of great puzzlement, it is of course an
immediate consequence of our extended principle of relativity and this
was eventually recognized by Poincaré (1854–1912) in 1899. With the
omission of the word 'dynamical', the principle of Newtonian
relativity now became the *first relativity postulate*, valid generally. It
reads:

> 'In any given space region there exists an infinite set of inertial
> frames in which the laws of physics take on a specially simple
> form. It is impossible to distinguish between two such frames of
> reference by any physical experiment.'

However, the acceptance of this postulate created a new puzzle,
namely that the velocity of light did not obey the velocity addition
law (1.17). It was Einstein (1879–1955) who recognized that this too
had to be accepted and that this led to a total re-analysis of the concepts of space and time and of the inter-relation of measurement and
observer. The resulting theory, the *special theory of relativity*, which is
based on Poincaré's first relativity postulate and Einstein's *second
relativity postulate* that

> 'every inertial observer measures the same value in free space of
> the speed of light relative to himself'.

will occupy us for most of the rest of this chapter. The word 'special'
in the title of the theory refers to the fact that it confines itself to
inertial observers. Later, Einstein developed the *general theory of
relativity*, which extends the theory to non-inertial observers, but we
shall not be concerned with this. We shall only be able to give a very
brief account even of the Special Theory, and the reader is referred
for a more detailed account to, for instance, Born, *Einstein's Theory of*

Relativity, or to D. Bohm, *The Special Theory of Relativity*, Benjamin, 1965.

Progress test

1.   The Michelson-Morley experiment showed that
   A   the velocity of light does not depend on the motion of the earth through the ether
   B   the velocity of light does not depend on the velocity of the light source
   C   the ether is not being dragged along by the earth
   D   the velocity of light is the same for all inertial observers.

Solutions

1.   (B)   The distance vector of separation of the two observers is vt and its sign depends on whether the vector is defined from O to O' or from O' to O.

2.   (C)   Velocity is not an invariant under a Galilean transformation.

3.   (C)   The components of a vector change under a rotation of axes.

2.6   Departures from Newtonian mechanics

Although Einstein was led to his theory through considerations relating to light, there were in fact at the time already observations on dynamical systems that were in conflict with Newtonian dynamics, and we shall further see that the predictions of Einstein's theory in the field of dynamics have been amply verified. The observations existing at the time related to the measurement of the momentum and velocity of fast electrons, and the first of them were due to Kaufmann in 1901, who measured the deflection of β-rays, which are electrons emitted in radioactive decays, in electric and magnetic fields. They showed that the ratio of momentum p to the product of mass and velocity mv, which according to (1.21) is constant and equal to unity, increased with velocity. The accompanying table shows the results of some of the early experiments, which indicate that the ratio becomes very large indeed as the electron velocity v approaches that of light c. This radical departure from Newtonian mechanics is, as we shall see, a natural consequence of Einstein's ideas, when applied to the motion of particles. It was the first indication that there might be an upper limit to the velocity of particles, and that this upper limit was the velocity of light. In this, it hinted at a possible unification of two, until then, quite separate branches of physics, mechanics and optics; a unification which was indeed achieved by the special theory of relativity.

v/c	p/mv
0·79	1·5
0·83	1·66
0·86	2·0
0·91	2·42
0·95	3·1

2.7   The relativistic addition law of velocities

Returning now to the considerations at the end of section 2.5, our first aim must clearly be to obtain a new velocity addition law in agreement with the second relativity postulate. Consider two observers O and O', and let O' have velocity v relative to O. Let both observe the motion of a particle P moving along the line OO', and let the respective measurements of the velocity of P be u and u'. Then we require there to be a 1–1 relationship between u and u', i.e., that, given u, we can uniquely

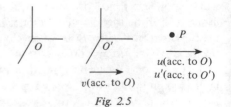

Fig. 2.5

determine u', and conversely. This is certainly the case for the Newtonian addition law, which gives

$$u' = u - v, \qquad \text{NR} \tag{2.8}$$

where the letters NR stand for non-relativistic, a word which has come
to be used for a theory which satisfies the first, but not the second
relativity postulate. This equation has a term in u, a term in u' and one
without either. Our task is now to find the simplest modification, which
still maintains the 1–1 relationship. Obviously, we must not add terms
in u^2 or higher powers, since this immediately spoils the 1–1 nature
of the relation. However, this is not so if we add a term in uu', and this
appears to be the simplest way to extend the structure of (2.8). Let us
try it. We then hope that the equation linking u and u' is of the form

$$puu' + qu + ru' + s = 0 \tag{2.9}$$

where p, q, r, s depend on v, but not on u or u'. By solving for u or u',
we easily see that the relation is 1–1. We now find the constants p, q, r, s
by treating three special cases for u and u':

(a) when $u' = 0$, $u = v$. This is a consequence of the definition of
relative velocity.

(b) When $u = 0$, $u' = -v$. This follows from the first postulate,
according to which O and O' are equivalent.

(c) When $u = c$, $u' = c$. This is the startling new consequence of the
second postulate, for the special case when P is a light pulse. It
goes totally against common sense and violates (2.8).

We now substitute the above into (2.9) and solve for the ratios $p{:}q{:}r{:}s$.
This gives

$$q = -\frac{1}{v}s, \qquad r = \frac{1}{v}s, \qquad p = -\frac{1}{c^2}s. \tag{2.10}$$

Substituting back into (2.9) and solving for u' we have the new
relativistic addition law of velocities

$$u' = \frac{u - v}{1 - \dfrac{uv}{c^2}}. \tag{2.11}$$

We have succeeded in arriving at a formula linking u and u', which
satisfies also the second relativity postulate. It does not of course
follow that it is correct, since it was the result of an inspired guess.
Only experiment can confirm its validity, and this has in fact happened.

We note that as $v/c \to 0$, the law (2.11) reduces to the non-
relativistic law (2.8). Hence Newtonian mechanics is the limiting case
of relativistic mechanics when the ratio of particle velocities to the

velocity of light tends to zero, or alternatively, when $c \rightarrow \infty$. Since deductions from Newton's laws have been abundantly verified under these limiting conditions, we would indeed expect that formulae of the theory of relativity must go over into classical (non-relativistic) formulae in the mathematical limit $c \rightarrow \infty$. This is an illustration of the *correspondence principle*, first formulated by Bohr in connection with quantum theory, which states that

> 'a new physical theory must contain within itself the theory which it supersedes in that limit in which the older theory had shown itself to be valid.'

Solution

1. (A) All the statements are correct, but only (A) follows from the Michelson-Morley experiment.

Progress test

1. If a body has velocity u relative to O and u' relative to O', and the velocity of O' relative to O is v, where v is parallel to u and u', then

A $\quad u' = \dfrac{u+v}{1 - \dfrac{uv}{c^2}}$

B $\quad u' = \dfrac{u-v}{1 - \dfrac{uv}{c^2}}$

C $\quad u' = \dfrac{v-u}{1 - \dfrac{uv}{c^2}}$

D $\quad u' = \dfrac{v-u}{1 + \dfrac{uv}{c^2}}$

2. An observer sees two particles moving in the same direction along a straight line with velocities u and v, where $u > v$. Their relative velocity is always

A $\quad = u - v$
B $\quad > u - v$
C $\quad < u - v$
D $\quad$ none of these.

2.8 Relativistic co-ordinate transformation

Consider the Galilean transformation $(2.5')$ for the case in which $v_x = v, v_y = 0, v_z = 0$. (In this and the subsequent sections, the relative velocity v between the two frames of reference will always be along the x-axis.) In this case, the addition law of velocities (1.17) can be written

$$\frac{dx'}{dt'} = \frac{dx}{dt} - v,$$

$$\frac{dy'}{dt'} = \frac{dy}{dt}, \qquad\qquad\qquad \text{NR (2.12)}$$

$$\frac{dz'}{dt'} = \frac{dz}{dt}.$$

On the assumption (2.6), that $t' = t$, we can at once write this in a directly integrable form,

$$dx' = dx - vdt, \qquad dy' = dy, \qquad dz' = dz \qquad \text{NR (2.12')}$$

and integrate it:

$$x' = x - vt, \qquad y' = y, \qquad z' = z, \qquad t' = t. \qquad \text{NR (2.13)}$$

The same cannot be done with the corresponding relativistic addition law of velocities (2.11),

$$\frac{dx'}{dt'} = \frac{\dfrac{dx}{dt} - v}{1 - \dfrac{v}{c^2}\dfrac{dx}{dt}} \qquad\qquad\qquad (2.14)$$

since, even if we put $t' = t$, the equation is not directly integrable. We therefore relax the condition $t' = t$, and instead look for some way of obtaining now two equations, for x' and for t', from (2.14). If we write (2.14) as

$$\frac{dx'}{dt'} = \frac{dx - vdt,}{dt - \dfrac{v}{c^2}dx} \qquad\qquad\qquad (2.15)$$

then we can obviously split this into two directly integrable equations,

$$dx' = \beta(dx - v\,dt)$$

$$dt' = \beta\left(dt - \frac{v}{c^2}\right)dx, \qquad\qquad\qquad (2.16)$$

where β does not depend on x, t, x', t', but may depend on v. It should be stressed that while (2.16) cannot be said to follow uniquely from (2.15), it does lead to linear equations in the co-ordinates, i.e., to the simplest form of a co-ordinate transformation compatible with (2.11),

$$x' = \beta(x - vt), \qquad\qquad\qquad (2.17)$$

$$t' = \beta\left(t - \frac{v}{c^2}x\right). \qquad\qquad\qquad (2.18)$$

To obtain β, we eliminate t, which leads to

$$x' + vt' = \beta\left(1 - \frac{v^2}{c^2}\right)x. \qquad (2.19)$$

But since O and O' are completely equivalent, it follows from (2.17) that

$$x = \beta(x' + vt'). \qquad (2.20)$$

Hence

$$\beta^2 = \left(1 - \frac{v^2}{c^2}\right)^{-1}, \qquad (2.21)$$

and we must take the positive root for β, since (2.17) must reduce to (2.13), as $c \to \infty$, by the corresponding principle. That the transformation is consistent can be shown by eliminating x between (2.17) and (2.18) and solving for t:

$$t = \beta\left(t' + \frac{v}{c^2}x'\right). \qquad (2.22)$$

Regarding y and z, it is simplest to assume that these are unchanged by the transformation. We therefore obtain the co-ordinate transformation for two frames of reference $Oxyz, O'x'y'z'$, with O' moving with velocity v relative to O in the x-direction:

$$x' = \frac{x - vt}{\sqrt{\left(1 - \frac{v^2}{c^2}\right)}}$$

$$y' = y$$

$$z' = z \qquad (2.23)$$

$$t' = \frac{t - \frac{v}{c^2}x}{\sqrt{\left(1 - \frac{v^2}{c^2}\right)}}$$

This, known as the *Lorentz transformation* after H. A. Lorentz (1853–1928) who discovered it, obviously reduces to the Galilean transformation (2.13) in the NR limit. We have therefore succeeded in obtaining a co-ordinate transformation that satisfies both the relativity postulates, but at the expense of giving up the universality of time. An important check on the validity of (2.23) will be found in worked example 2.1.

    Finally, it should be noted that velocities transverse to the relative velocity of the two observers are altered by the transformation.

Solutions

1. (B) This is straight bookwork.

2. (B) This is an application of the formula of question 1 to the case where u and v are in the same direction so that $1 - uv/c^2 < 1$.

Thus

$$\frac{dy'}{dt'} = \frac{\sqrt{\left(1 - \dfrac{v^2}{c^2}\right)}}{1 - \dfrac{v}{c^2}\dfrac{dx}{dt}}\frac{dy}{dt}. \tag{2.24}$$

    The Lorentz transformations form the basis of the mathematical treatment of the special theory of relativity. With their help it is possible to investigate the nature of time and space, and to show how it differs from that based on our commonsense notions. Only by suspending commonsense and using the formalism of mathematics can we do this. A glimpse of the power of this method is given in worked example 2.2.

Worked example 2.1. A light signal is emitted by observer O at exactly the moment when observer O' passes O with velocity v relative to him. Show that the velocity of light in any direction is the same, as measured by both observers.

    According to O, let the signal reach a point $P(x,y,z)$ at time t. Then

$$OP^2 = x^2 + y^2 + z^2 = c^2 t^2. \tag{1}$$

Now according to O', the signal is then at $P(x',y',z')$. From (2.23) we have

$$O'P'^2 = x'^2 + y'^2 + z'^2$$

$$= \frac{(x - vt)^2}{1 - v^2/c^2} + y^2 + z^2$$

$$= \frac{(x - vt)^2}{1 - v^2/c^2} - x^2 + c^2 t^2, \text{ using (1)}$$

$$= \frac{[ct - (vx/c)]^2}{1 - v^2/c^2}$$

$$= c^2 t'^2, \text{ using (2.23) again.}$$

Thus O' measures the same velocity of light as O.

    It should be noted that this demonstration of the agreement of the Lorentz transformation with the second relativity postulate shows the correctness of the transformation $y' = y$, $z' = z$, which had not previously been verified.

Worked example 2.2. Show that if O observes two events at (x_1, y_1, z_1) and (x_2, y_2, z_2) simultaneously at time t, then to O' moving with velocity v relative to O, one event appears prior to the other.

Let O' observe the events at

$$(x'_1, y'_1, z'_1, t'_1), \quad (x'_2, y'_2, z'_2, t'_2).$$

Then

$$x'_1 = \beta(x_1 - vt)$$

$$x'_2 = \beta(x_2 - vt)$$

$$\overline{x'_1 - x'_2 = \beta(x_1 - x_2)}$$

$$t'_1 = \beta\left(t - \frac{v}{c^2}x_1\right)$$

$$t'_2 = \beta\left(t - \frac{v}{c^2}x_2\right)$$

$$\overline{t'_1 - t'_2 = \frac{\beta v}{c^2}(x_2 - x_1)}$$

$$= \frac{v}{c^2}(x'_2 - x'_1).$$

Thus for $v > 0$, the event nearer to O' in space will appear to have happened later. We conclude that simultaneity is a relative concept. Further, if a second observer moves in a direction opposite to O', relative to O, then to him the events will appear to have happened in the opposite order. Note, however, that for all observers the distance between the events is

$$\Delta x = \frac{c^2}{v}\Delta t > c\,\Delta t$$

since $v < c$. Hence the distance is such that no signal could have passed between the two events, so that they must be causally independent. The fact that their time-ordering depends on the observer does not therefore upset our ideas of cause and effect.

It can also be shown that if two events are such that $\Delta x < c\,\Delta t$, then the time-order of the events is the same for all observers, and they cannot appear simultaneous to any.

Progress test

1. The Lorentz transformation must be used to correlate the measurements of two observers, when which of the following is very large?

   A   The distance between the observers
   B   The relative velocity of the observers
   C   The relative acceleration of the observers
   D   The velocity of light.

2.9   The relativistic concept of time

We have so far developed the theory entirely on an algebraic basis and been led to the conclusion that time is different for two observers moving relative to each other. This quite astonishing conclusion is in total conflict with commonsense experience, expressed in Newton's famous statement about 'absolute, true and mathematical time', quoted at the beginning of this chapter. It is not therefore surprising that some of the conclusions at which we shall arrive also conflict with common-sense experience. It must always be remembered that this conflict is only significant when the relative speeds of the observers are of the order of the speed of light, a condition quite outside our experience.

Let us illustrate these considerations by means of an imagined experiment (a *Gedanken experiment*,* which can be thought of, but not actually performed). Consider two observers, each with a long mirror, moving parallel to the mirrors distance l apart with velocity v. Each now uses as the basis of his timing mechanism the time taken for a light pulse to travel to the other's mirror and back. Since each considers himself at rest and the other moving, each thinks that the time for one tick on his clock is $2l/c$, while the time for one tick on the other's clock is

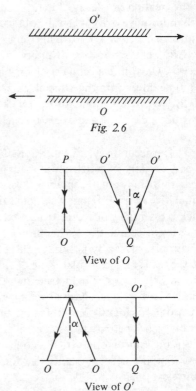

Fig. 2.6

Fig. 2.7

$$\frac{2l}{c\cos\alpha} = \frac{2l}{c\sqrt{1 - v^2/c^2}}.$$

Thus, each concludes that the other's clock is slow, i.e., a clock moving with velocity v relative to an observer loses time according to him by a factor $\sqrt{1 - v^2/c^2}$ compared with his own clock, which is stationary relative to him. This phenomenon is called *time dilation*. (See also problem 2.6.)

Time dilation has been observed experimentally. A mu-meson, which is one of the unstable particles created in high-energy nuclear reactions, has a measured lifetime of about 2×10^{-6} sec. Mu-mesons that are created by cosmic-ray bombardment in the upper atmosphere at heights of about 30 km with speeds approaching that of light have been found to reach ground level, although even at the speed of light this distance takes 10^{-4} s to travel. This is possible because according to the observer on the ground, who has measured the distance of 30 km, the mu-meson clock ticks much more slowly than his own — sufficiently slowly to lengthen the mu-meson lifetime by a factor of 100. An observer travelling with the mu-meson would of course measure the mu-meson lifetime as 10^{-6} s, but as we shall see below, he will measure the distance to the ground as 100 times less!

Another consequence of time dilation is the so-called *twin paradox*, according to which a traveller who goes away and returns ages less than his stay-at-home twin. This has indeed been verified for elementary particles. The explanation of the apparent paradox is that the twins are not equivalent observers, since if the stay-at-home is an inertial observer,

* This is the German technical term for 'imagined experiment', which is frequently used also in the English literature.

the traveller, who must undergo an acceleration at some stage if he is to return, is not. [See J. Bronowski, 'The Clock Paradox', *SA* 291, February 1963, p. 134.]

Although all inertial observers are dynamically equivalent with regard to the measurement of the time interval between two events, we single out for special attention the observer in whose frame of reference the space co-ordinates of the two events are the same. [If the events are the creation and decay of the mu-meson, and the mu-meson is travelling with uniform velocity, then the observer travelling with the mu-meson is such an observer.] We call his frame of reference the *rest frame* for the two events, and the time interval which he measures the *proper time interval*. Only in this frame is it possible to record the beginning and end of a time interval on the same clock. Other observers, in motion relative to the rest frame for the two events, measure *improper time intervals* and for this require two synchronized clocks in different places

The accurate synchronization of clocks in different places is not an easy matter. To synchronize them in the same place and then to separate them is not acceptable, since even the smallest error is cumulative in time. [It may be noted that, as a measuring device, a clock is essentially different from a measuring rod. A small error in the latter is not cumulative, either in space or time.] For that reason we synchronize clocks by passing signals between them, and in practice we use the fastest signal available, i.e., light. But this is only possible if we know the speed of light accurately, and to measure the speed of light we have to time a pulse over a measured distance, i.e., we need synchronized clocks! [Measuring the time of a round trip, using mirrors, is not permissible, since we have no guarantee that the speed of light is the same in opposite directions.] It was considerations such as these that, in the first instance, led Einstein to reject the whole idea of a universal time.

Solution

1. (B) The theory of relativity must be used when observer velocities approach the velocity of light.

Progress test

1. Which one of the following concepts of Newtonian physics is still valid in relativity physics?

    A   Absolute space and time
    B   Physical measurement is independent of the observer
    C   Simultaneity of two events
    D   Cause and effect relation between two events.

2. The proper time interval is the interval between two events as measured by an inertial observer.

    A   who is in an inertial frame of reference
    B   who is at rest relative to both events

C who is present at both events

D who uses a clock at rest relative to himself.

2.10 The relativistic concept of length

The relativity of time measurement leads to a corresponding relativity of length measurement. Consider a measuring rod of length s at rest relative to O, and let O' measure its length by measuring the time t' it takes him to travel along the rod with velocity v relative to O. Then we are considering two events—O' passing the beginning and passing the end of the rod—and O' measures the proper time between these events. If O now measures the time t which he considers O' to have taken, then it follows from the last section that because of time dilation, $t > t'$ and in fact

$$t' = t\sqrt{1 - v^2/c^2}. \tag{2.25}$$

But the length of the rod as measured by O and O' respectively is

$$s - vt \quad \text{and} \quad s' = vt'. \tag{2.26}$$

Hence

$$s' = s\sqrt{1 - v^2/c^2}, \tag{2.27}$$

i.e., the measuring rod appears shorter to an observer moving relative to it. This phenomenon is called the *length contraction*, and the length of the rod as measured in its own rest frame, i.e., by O, is called its *proper length*.

If we now remember the observer travelling with the mu-meson, it is clear that for him the height of the atmosphere is contracted in exactly the same ratio as the mu-meson lifetime was extended to the observer on the ground.

Progress test

1. If an observer moves relative to a clock and a ruler, he finds that

A the clock is slow and the ruler shortened

B the clock is slow and the ruler lengthened

C the clock is fast and the ruler shortened

D the clock is fast and the ruler lengthened.

2.11 Relativistic dynamics

We have already noted the experimental fact (see section 2.6) that the mass of a body, as measured by the ratio of its momentum to its velocity, increases with the velocity. Later, more accurate experiments verified that the momental mass of a body moving with velocity v is given by the formula

$$m(v) = \frac{m_0}{\sqrt{1 - v^2/c^2}},$$

where m_0 is the mass of the body at rest.

We shall now verify, by means of another *Gedanken experiment*, that this formula is in accord with the results of the theory of relativity obtained so far. In this experiment, we shall continue to assume the validity of the laws of conservation of energy and momentum, but because of the above formula, we shall suspend judgment regarding the law of conservation of mass.

Let two observers O and O', moving with velocity v relative to each other, aim two identical perfectly elastic particles at each other along a line perpendicular to their line of motion, at the moment when they pass each other. Let the components of velocity of the particle aimed by O be $(0, w)$ when measured by O and (u', w') when measured by O', and conversely. Then it follows from (2.24) that

$$w' = \sqrt{1 - v^2/c^2}\, w. \tag{2.28}$$

Because of conservation of energy, the velocity of rebound equals the velocity of approach and so, after the collision, the velocities merely have their signs reversed. Finally, let O measure the masses of the two particles as m and m'. Then, as he applies conservation of momentum along the line of w, he obtains

$$mw - m'w' = -mw + m'w', \tag{2.29}$$

that is,

$$m' = \frac{m}{\sqrt{1 - v^2/c^2}}. \tag{2.30}$$

Now, by making w very small, O can ensure that the motion of his particle in his frame of reference is given by Newtonian mechanics, and so he can take for m the usual Newtonian value, defined by (1.21). Since it is measured in his rest frame, it is called the *rest mass* of the particle and denoted by m_0. The other particle being identical, also has a rest mass m_0, but when it moves with velocity v relative to O, then its mass, as measured by the ratio of its momentum to its velocity, is

$$m(v) = \frac{m_0}{\sqrt{1 - v^2/c^2}}. \tag{2.31}$$

Solution

1. (A) This is the effect of time dilation and length contraction.

Solutions

1. (D) See worked example 2.2.

2. (C) This is the definition of proper time. Under certain circumstances the observer in (B) would also measure proper time.

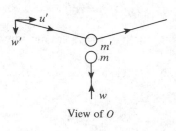

View of O

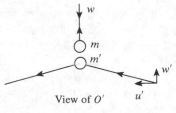

View of O'

Fig. 2.8

This confirms the formula which, as was stated at the beginning of this section, has been verified experimentally. The formula also verifies the experimental fact that particle velocities never reach the velocity of light.

If we write the mass correction as

$$\Delta m = m - m_0 = m_0 \left\{ \left(1 - \frac{v^2}{c^2} \right)^{-1/2} - 1 \right\}, \qquad (2.32)$$

where we now put m for $m(v)$, then in the NR limit, we can expand this and obtain

$$\Delta m = m_0 \left(1 + \frac{v^2}{2c^2} + \cdots - 1 \right) \simeq \tfrac{1}{2} m_0 \frac{v^2}{c^2} \qquad (2.33)$$

or

$$mc^2 = m_0 c^2 + \tfrac{1}{2} m_0 v^2. \qquad \text{NR} \quad (2.34)$$

Dimensionally this is an energy equation, i.e., each term has the dimensions of energy, with the obvious interpretation

total energy = rest-mass energy + kinetic energy   (2.35)

The rest-mass energy owes its name to the facts that it exists even when a particle is at rest, and that the only property of the particle on which it depends is its mass. We therefore are led to the suggestion that mass is a form of energy and that the law of conservation of matter must be abandoned. As physics has developed, the concepts of mass and matter have been progressively refined and further and further removed from common sense. To Newton, the mass of a body was 'the quantity of matter in the body', but we now treat matter, also called rest mass, as a form of energy, while inertia is a consequence of the momental mass, which is a function of kinetic energy. Finally, quantum mechanics has shown that matter can appear in the form of waves as well as particles! [E. Schrödinger, 'What is matter', SA 241, September 1953, p. 52.]

As we have postulated that the law of conservation of energy is still valid in relativistic dynamics, we conclude that (2.35) is generally valid, although the expression (2.34) yields a formula for the kinetic energy that is valid only non-relativistically. Thus we conclude that, for all velocities,

rest-mass energy $E_0 = m_0 c^2,$ (2.36)

total energy $E = mc^2 = \dfrac{m_0 c^2}{\sqrt{1 - v^2/c^2}}$ (2.37) and hence that

kinetic energy $T = mc^2 - m_0 c^2$

$$= m_0 c^2 \left\{ \frac{1}{\sqrt{1 - v^2/c^2}} - 1 \right\}. \qquad (2.38)$$

It cannot be stressed too strongly that there is no way in which the relativistic expression for T can be written down more simply than in (2.38). It is easily verified that the non-relativistic expression

$$T = \tfrac{1}{2} m_0 v^2 \qquad \text{NR} \qquad (2.39)$$

is the limit of (2.38) as $c \to \infty$.

The expression (2.38) has recently been directly verified by W. Bertozzi. [*Amer. J. Phys.* **32**, 551 (1964)] in an experiment in which the velocity of electrons in a beam was measured by timing them over a path of known length, and their kinetic energy was determined by absorbing them in a calorimeter. The results clearly differ radically from the Newtonian prediction.

Finally, from (2.31), we have for the relativistic momentum

$$p = \frac{m_0 v}{\sqrt{1 - v^2/c^2}} \qquad (2.40)$$

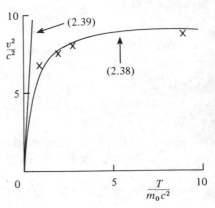

Fig. 2.9

and eliminating v between (2.37) and (2.40) we have

$$E^2 = p^2 c^2 + m_0^2 c^4. \qquad (2.41)$$

Expressions (2.37), (2.40) and (2.41) are the basis for the dynamics of particles moving with speeds close to that of light.

Worked example 2.3. A particle of rest mass m_0, moving at speed $v = \tfrac{3}{5} c$ collides inelastically with a similar particle at rest. Find the speed and rest mass of the composite particle.

Let the speed and rest mass of the composite particle be V and M_0 respectively. Then, using conservation of momentum and energy,

$$\frac{m_0 v}{\sqrt{1 - v^2/c^2}} = \frac{M_0 V}{\sqrt{1 - V^2/c^2}}$$

$$\frac{m_0 c^2}{\sqrt{1 - v^2/c^2}} + m_0 c^2 = \frac{M_0 c^2}{\sqrt{1 - V^2/c^2}}.$$

Hence

$$\frac{v}{1 + \sqrt{1 - v^2/c^2}} = V$$

Putting $v = \frac{3}{5}c$ gives $V = \frac{1}{3}c$.

$$\therefore \quad M_0 = \frac{\frac{3}{5}\sqrt{\frac{8}{9}}}{\frac{1}{3}\sqrt{\frac{16}{25}}} m_0 = \frac{3\sqrt{2}}{2} m_0 = 2.12 \, m_0.$$

Progress test

1. The ratio of momentum to velocity is given by

    A $\dfrac{m_0}{1 - v^2/c^2}$

    B $m_0(1 - v^2/c^2)$

    C $\dfrac{m_0}{\sqrt{1 - v^2/c^2}}$

    D $m_0\sqrt{1 - v^2/c^2}$.

2. A particle has rest mass m and velocity v. If $m = m_0/\sqrt{1 - v^2/c^2}$, the kinetic energy of the particle is

    A mc^2
    B $\frac{1}{2}m_0 v^2$
    C $\frac{1}{2}mv^2$
    D $mc^2 - m_0 c^2$

3. An atomic nucleus undergoes spontaneous fission into two fragments. Which one of the following statements is incorrect?

    A   The total rest mass of the fission fragments is less than that of the original nucleus
    B   The total momentum of the fission fragments is equal to that of the original nucleus
    C   The total kinetic energy of the fission fragments is greater than that of the original nucleus
    D   The total energy of the fission fragments is less than that of the original nucleus.

2.12. The origin of the magnetic field

For our final application of the special theory of relativity, we investigate the magnetic effect due to a steady electric current. [J. M. Osborne, *School Science Review* **45**, 54, 1963.]

From our point of view, what is a steady current to one observer, is charge at rest to another observer, moving with the current. And charges at rest do not produce a magnetic field! Let us assume that the current in a conducting wire is due to the motion of negative charges in the conductor. Since the conductor as a whole is neutral, there must be an equal and opposite amount of positive charge, which we assume to be at rest in the conductor. This is reasonable, since the negative charge is carried by the electrons, which are light, while the positive charge is located on the ions that are massive and make up the structure of the conductor. We now consider the force between two parallel wires that carry equal currents in the same direction. Let there be n positive charges of magnitude e per unit length of wire. Then the positive charge A is repelled by the positive charge B and attracted by the negative charge C. We shall show that these effects, when taken over the whole of the two wires, do not cancel.

The force dF_+ on the positive charge at A due to the positive charges in the element dx is, from Coulomb's law,

$$dF_+ = -\frac{ne^2\,dx}{4\pi\epsilon_0\,r^2} \qquad (2.42)$$

along r. The negative sign indicates that it is a repulsion. The corresponding force dF_- due to the negative charges is found as follows. If these are moving with velocity v then, to the observer at A, the element dx suffers a length contraction, i.e., the negative charges in dx will appear to A to occupy an element of length

$$dx\sqrt{1 - v^2/c^2}.$$

Hence to A, the density of negative charges in the other wire appears higher than that of the positive charges, and the force dF_- is therefore greater in magnitude than dF_+, that is,

$$dF_- = \frac{ne^2\,dx}{4\pi\epsilon_0\,r^2(1 - v^2/c^2)^{1/2}} \qquad (2.43)$$

The resultant force directed towards A is

$$dF = dF_+ + dF_-$$
$$= \frac{ne^2\,dx}{4\pi\epsilon_0\,r^2}\left[\left(1 - \frac{v^2}{c^2}\right)^{-1/2} - 1\right]$$
$$= \frac{ne^2\,v^2\,dx}{8\pi\epsilon_0\,c^2\,r^2}, \qquad (2.44)$$

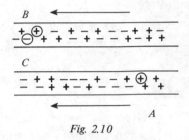

Fig. 2.10

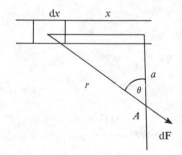

Fig. 2.11

where we have used the binomial theorem, as in (2.33). The velocities of the electrons are in fact experimentally found to be of the order of less than 1 m s^{-1}, so that the approximation is extremely good.

When we consider the total force **F** due to the whole wire, then it is clear that only the component of d**F** perpendicular to the wire contributes. Hence the total force on A is

$$\mathbf{F} = \int_{-\infty}^{\infty} \frac{ne^2 \, v^2 \, dx \cos \theta}{8\pi\epsilon_0 \, c^2 \, r^2}, \tag{2.45}$$

where θ is the angle between r and the perpendicular from A to the wire. If the distance between the wires is a, then

$$x = a \tan \theta, \qquad dx = a \sec^2 \theta \, d\theta, \qquad r = a \sec \theta,$$

and, in terms of θ, the integration is from $-\tfrac{1}{2}\pi$ to $+\tfrac{1}{2}\pi$. Hence

$$\mathbf{F} = \frac{ne^2 \, v^2}{8\pi\epsilon_0 \, c^2 \, a} \int_{-\pi/2}^{\pi/2} \cos \theta \, d\theta$$

$$= \frac{n^2 \, e^2 \, v^2}{4\pi\epsilon_0 \, c^2 \, a}. \tag{2.46}$$

Similarly, there is an equal attractive force between the wires due to the current in the lower wire on the positive charges in the upper wire. The current itself is given by

$$i = nev. \tag{2.47}$$

Hence the resultant force of attraction between two wires, each carrying a current i, is

$$\frac{i^2}{2\pi\epsilon_0 \, c^2 \, a}. \tag{2.48}$$

Now the electric permittivity ϵ_0 and the magnetic permeability μ_0, both in free space are related through the equation [see Jackson, chapter 1.]

$$\epsilon_0 \mu_0 = 1/c^2. \tag{2.49}$$

Hence finally, the force of attraction is

$$\frac{\mu_0 \, i^2}{2\pi a}. \tag{2.50}$$

This in electrodynamics is known as Ampére's law, and is usually taken as the fundamental law of the subject. We see that the special theory of relativity unites electrostatics and electrodynamics, and makes Ampére's law a consequence of Coulomb's law. What is particularly remarkable about this derivation is that it is valid for all velocities of the negative charges, however small; in other words, we have found an effect resulting from the special theory of relativity which is measurable at all velocities, however small.

2.13 Problems

2.1. Show that the distance between two points is invariant under a Galilean transformation, but not under a Lorentz transformation. [Hint: the distance between two points at r_1 and r_2 is $|r_1 - r_2|$.]

2.2. A man holding a parcel of mass m is standing in a lift which is being accelerated upward by a constant force F. The total mass of lift plus passenger is M.

(a)   What is the acceleration of the lift?

(b)   What weight does the parcel appear to have for the passenger?

(c)   If the passenger drops the parcel from height h, how long does it take to reach the floor?

(d)   If the lift changes to uniform velocity, what is the value of F?

2.3. In the Michelson–Morley experiment, a beam was split by a half-silvered mirror M into two components, which then travelled through paths of identical length l to mirrors M_1 and M_2 and back to M, before being re-united and observed in the telescope T. If the earth moves with velocity v relative to the ether along MM_1, show that the time difference for the two components of the beam to reach T is

$$\Delta t = \frac{l}{c}\frac{v^2}{c^2}$$

to order v^2/c^2. If $v = 30$ km s^{-1} (velocity of earth relative to sun) and $l = 15$ m, show that the corresponding path difference is of the order of the wavelength of visible light.

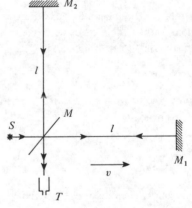

Fig. 2.12

2.4. Use (2.11) to verify the important relation

$$\left(1 - \frac{u'^2}{c^2}\right)\left(1 - \frac{uv}{c^2}\right)^2 = \left(1 - \frac{u^2}{c^2}\right)\left(1 - \frac{v^2}{c^2}\right).$$

2.5 In a laboratory experiment, the speeds of two electrons moving in opposite directions are each found to be $0.8c$. What is their relative velocity?

2.6. By writing down the Lorentz transformation for two different
points at the same time, as measured by one observer, derive the
formula for the length contraction (2.27). Similarly derive the
time-dilation formula.

2.7. By considering the concepts of time dilation and length contraction,
show that for any two events the proper time interval is the shortest
time interval and the proper length the longest length between the
two events, as measured by any inertial observer.

2.8. Plot on the same graph the kinetic energy according to (2.38) and
(2.39) for particle velocities from $v = 0$ to $v = 0.9c$. If measurements
of velocity and kinetic energy can be made with an accuracy of
one per cent, up to what velocity would you judge the NR
formula to be valid?

2.9. A particle of rest mass M breaks up spontaneously into two equal
particles. If, after the break-up, each particle moves with speed
$\frac{3}{5}c$, find

(a)   the rest mass of each of the particles
(b)   the kinetic energy of each of the particles.

Verify that the total energy of the system is conserved.

(The conversion of mass energy into kinetic energy in this process
is the basis of the production of nuclear energy through fission.)

2.10. Two observers, moving with relative velocity v, measure the total
energy and momentum of a particle moving parallel to their
relative velocity as E,p and E',p' respectively. Use the formula in
example 2.4 to show that

$$E' = \frac{E - pv}{\sqrt{1 - v^2/c^2}}, \qquad p' = \frac{p - Ev/c^2}{\sqrt{1 - v^2/c^2}}.$$

3. Central forces

3.1 Motion of a point in a plane

The basic forces of nature, such as the gravitational and electric ones, act between particles, and their line of action is along the line joining the particles. In general, both particles may be expected to move in a given reference frame under their mutual forces, but when one is very much more massive than the other, then the frame of reference in which it is at rest approximates to an inertial reference frame for the other particle. In that case we may take the force on the other particle, say at P, to be directed towards a fixed point in that frame of reference. Even when this is not the case, a small modification (see section 3.9) enables us to treat the motion of both particles as due to forces directed towards a fixed point. There are other situations, such as the one of an atom displaced from its mean position in a crystal lattice, where the force is the resultant of a number of forces that is directed towards a fixed point, in this case the undisplaced position of the atom. Such forces, that are directed towards a fixed point, are called *central forces* and they play an important role in our description of the motion of particles. We shall show that motion under central forces always takes place in a plane, and the mathematical formalism which we shall use to describe such a motion is that of polar co-ordinates in the plane.

The polar co-ordinates of a point P with position vector $\mathbf{r} = \overrightarrow{OP}$ relative to the origin O of a reference frame are the distance $r = OP$ and the angle θ, which $\mathbf{r}$ makes with a fixed line OX in the reference frame. We now define unit vectors $\hat{\mathbf{a}}$ and $\hat{\mathbf{b}}$ at P,* along and perpendicular to $\overrightarrow{OP}$. As the point P moves in the plane, defined by O, X and P, these are constant in magnitude, but not in direction. We next draw the vectors on a unit circle for two polar angles θ and $\theta + d\theta$. Then it is clear that the change in $\hat{\mathbf{a}}$, $d\hat{\mathbf{a}}$, is of magnitude $d\theta$ and, in the limit as $d\theta \rightarrow 0$, perpendicular to $\hat{\mathbf{a}}$. Hence $d\hat{\mathbf{a}}$ is in direction $\hat{\mathbf{b}}$. Similarly, $d\hat{\mathbf{b}}$ has magnitude $d\theta$ and is in direction $-\hat{\mathbf{a}}$. Since $\hat{\mathbf{a}}$ and $\hat{\mathbf{b}}$ are unit vectors, we have

$$d\hat{\mathbf{a}} = \hat{\mathbf{b}}\,d\theta \quad \text{and} \quad d\hat{\mathbf{b}} = -\hat{\mathbf{a}}\,d\theta. \tag{3.1}$$

* Vectors with hats always denote unit vectors.

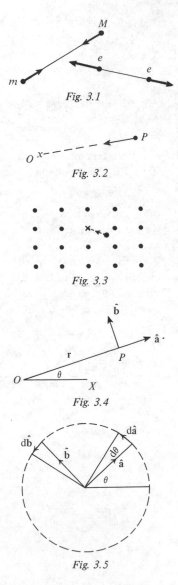

Fig. 3.1

Fig. 3.2

Fig. 3.3

Fig. 3.4

Fig. 3.5

Dividing by dt, we find the rate of change of **a** and **b** in terms of the angular velocity $\dot{\theta}$ of the radius vector **r**,

$$\dot{\hat{\mathbf{a}}} = \hat{\mathbf{b}}\dot{\theta}, \qquad \dot{\hat{\mathbf{b}}} = -\hat{\mathbf{a}}\dot{\theta}. \tag{3.2}$$

Now the position of P is completely determined by the radius vector

$$\mathbf{r} = r\hat{\mathbf{a}}. \tag{3.3}$$

To find the velocity and acceleration of P, we differentiate (3.3) and make use of (3.2):

$$\mathbf{v} = \dot{\mathbf{r}} = \dot{r}\hat{\mathbf{a}} + r\dot{\hat{\mathbf{a}}},$$

that is,

$$\mathbf{v} = \dot{r}\hat{\mathbf{a}} + r\dot{\theta}\hat{\mathbf{b}}; \tag{3.4}$$

$$\mathbf{a} = \ddot{\mathbf{r}} = \ddot{r}\hat{\mathbf{a}} + \dot{r}\dot{\hat{\mathbf{a}}} + \dot{r}\dot{\theta}\hat{\mathbf{b}} + r\ddot{\theta}\hat{\mathbf{b}} + r\dot{\theta}\dot{\hat{\mathbf{b}}},$$

that is,

$$\mathbf{a} = (\ddot{r} - r\dot{\theta}^2)\hat{\mathbf{a}} + (2\dot{r}\dot{\theta} + r\ddot{\theta})\hat{\mathbf{b}}. \tag{3.5}$$

[Please note that the acceleration vector **a** must not be confused with the unit vector $\hat{\mathbf{a}}$. Also, in the notation employed here, $|\ddot{\mathbf{r}}|$ is not the same as $\ddot{r}$. In $|\ddot{\mathbf{r}}|$ we first differentiate **r** twice and then take the modulus, while in $\ddot{r}$ we first take the modulus of **r** and then differentiate twice.]

It is easily seen that the coefficient of $\hat{\mathbf{b}}$ in this equation can be written

$$2\dot{r}\dot{\theta} + r\ddot{\theta} = \frac{1}{r}\frac{d}{dt}(r^2\dot{\theta}), \tag{3.6}$$

so that we have finally

$$\mathbf{a} = (\ddot{r} - r\dot{\theta}^2)\hat{\mathbf{a}} + \frac{1}{r}\frac{d}{dt}(r^2\dot{\theta})\hat{\mathbf{b}}. \tag{3.7}$$

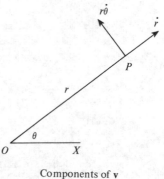

Components of **v**

Fig. 3.6

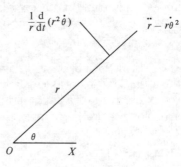

Components of **a**

Fig. 3.7

Equations (3.3), (3.4) and (3.7) give us the components, in polar coordinates, of the position, velocity and acceleration vectors of a point moving in a plane.

Progress test

1. The polar co-ordinates of a point are $r = 4, \theta = 210°$. Its Cartesian co-ordinates are

    A $(2, 2\sqrt{3})$

    B $(2\sqrt{3}, 2)$

    C $(-2, 2\sqrt{3})$

    D $(-2, -2\sqrt{3})$.

2. Which one of the following expressions is incorrect?

    A $\hat{\mathbf{a}} \cdot \hat{\mathbf{a}} = 1$

    B $\dot{\hat{\mathbf{a}}} \cdot \hat{\mathbf{a}} = 0$

    C $\hat{\mathbf{a}} \cdot \hat{\mathbf{b}} = 0$

    D $\dot{\hat{\mathbf{a}}} \cdot \hat{\mathbf{b}} = \hat{\mathbf{a}} \cdot \dot{\hat{\mathbf{b}}}$

3. The magnitude of the radial velocity vector $\dot{\mathbf{r}}$ is

    A $\dot{r}$

    B $|\dot{\mathbf{r}} - \mathbf{r}\dot{\theta}|^2$

    C $\sqrt{(\dot{r}^2 + r^2\,\dot{\theta}^2)}$

    D $\sqrt{(\dot{r}^2 + 2r\dot{r}\dot{\theta} + r^2\,\dot{\theta}^2)}$

3.2   Central conservative forces

Forces are in general functions of position, by which we mean that both the magnitude and direction of the force $\mathbf{F}$ which acts on a particle when it is at the point P, position vector $\mathbf{r}$, depends on the position of this point. In other words, $\mathbf{F}$ is a function of $\mathbf{r}$, or alternatively of r and θ. Hence we can write

$$\mathbf{F} = \mathbf{F}(\mathbf{r}) = \mathbf{F}(r, \theta). \tag{3.8}$$

Now we must remember that the direction of $\mathbf{F}$ is along the line OP, but that the direction of the reference line OX is quite arbitrary. It is in no way related to the physical situation, and so the magnitude and direction of a central force cannot depend on θ. Hence for a central force, we can put

$$\mathbf{F} = \hat{\mathbf{a}}F(r), \tag{3.9}$$

where we have used the unit vector $\hat{\mathbf{a}}$ to describe the direction of $\mathbf{F}$, while the scalar function $F(r)$ describes the magnitude. We can always integrate such a function and thus get another function, i.e., we can integrate $F(r)$ to obtain another function, say, $- V(r)$. In other words, $F(r)$ can be written as the negative gradient of another function,

$$F(r) = - \frac{\mathrm{d}V}{\mathrm{d}r}, \tag{3.10}$$

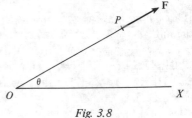

Fig. 3.8

so that (see section 1.10) a central force is a conservative force and $V(r)$ is the potential energy.

On using Newton's second law, and substituting (3.9) into (3.7), we have

$$\mathbf{F}(r) = \hat{\mathbf{a}} F(r) = m\ddot{\mathbf{r}}, \tag{3.11}$$

that is,

$$F(r) = m(\ddot{r} - r\dot{\theta}^2), \tag{3.12a}$$

$$\frac{d}{dt}(r^2 \dot{\theta}) = 0. \tag{3.12b}$$

Equations (3.12a) and (3.12b) are the equations of motion of a particle P under the central force $F(r)$, i.e., if we eliminate the time t between them, we obtain an equation in r and θ, which defines the path of the particle in the plane described under the force $F(r)$.

Progress test

1. An atom displaced from its mean position in a crystal lattice is under a force which is always directed towards the mean position and is proportional to its distance from it. The force is therefore of the form (K is a positive constant)

   A $Kr\hat{\mathbf{a}}$

   B $-Kr\hat{\mathbf{a}}$

   C $Kr\hat{\mathbf{b}}$

   D $-Kr\hat{\mathbf{b}}$.

2. The force in question 1 (V_0 is a constant)

   A   can be derived from a potential $V_0 + \frac{1}{2}Kr^2$

   B   can be derived from a potential $V_0 - \frac{1}{2}Kr^2$

   C   can be derived from a potential $V_0 + Kr^2$

   D   cannot be derived from a potential.

3. Which one of the following properties does the force in question 1 not possess?

   A   It is central
   B   It is conservative
   C   It depends on distance
   D   It is repulsive.

3.3 Angular momentum

In the previous section we have met vectors which were functions of the position of a point P, i.e., they were functions of the position vector $\mathbf{r} = \overrightarrow{OP}$. We can think of such a vector, say $\mathbf{s}$, at P turning the vector $\mathbf{r}$ about O, and may wish to obtain a precise definition of what might be called the turning effect. For instance, $\mathbf{s}$ might be a force acting at the end of a handle OP and turning it about O.

Quite generally, the vectors $\mathbf{r}$ and $\mathbf{s}$ define a plane and we can draw a perpendicular OZ to that plane. Only the component of $\mathbf{s}$, which is perpendicular to $\mathbf{r}$, has a turning effect about OZ and so it is natural to define the turning effect of $\mathbf{s}$ about O as

$$sr \sin \theta$$

where θ is the angle between $\mathbf{r}$ and $\mathbf{s}$. We are thus led to a quantity which has a magnitude, $rs \sin \theta$ and a direction $\overrightarrow{OZ}$. This defines a vector and, as this new vector is made up of the two vectors $\mathbf{r}$ and $\mathbf{s}$, we call it the *vector product* of $\mathbf{r}$ and $\mathbf{s}$ and write it $\mathbf{r} \times \mathbf{s}$. This is therefore a vector of magnitude $rs \sin \theta$ and in a direction which is perpendicular to the plane defined by $\mathbf{r}$ and $\mathbf{s}$. To distinguish between the two possible directions of the perpendicular, we use the *right-hand screw convention*, i.e., in turning a right-handed screw from $\mathbf{r}$ to $\mathbf{s}$ we move it along the direction $\mathbf{r} \times \mathbf{s}$.

Although we have spoken of the 'direction' of $\mathbf{r} \times \mathbf{s}$, the meaning of this word is here clearly quite different from the one it has when we speak of the 'direction' of the vector $\mathbf{s}$. The latter may be a force or a momentum and is physically associated with motion along that direction. The vector product on the other hand is physically associated with motion *about* a direction. Another peculiarity of the vector product is that it is *non-commutative*, i.e., it changes in value, if the order of the factors is reversed. It is easily seen that

$$\mathbf{r} \times \mathbf{s} = -\mathbf{s} \times \mathbf{r}. \tag{3.13}$$

After these mathematical preliminaries, we now use the concept of the vector product to consider the turning effect of the momentum vector $\mathbf{p} = m\mathbf{v}$ of the momentum of a particle of mass m at P moving with velocity $\mathbf{v}$. This turning effect is called the *angular momentum* of the particle about O and defined as

$$\mathbf{l} = \mathbf{r} \times \mathbf{p}. \tag{3.14}$$

From now on, we shall refer to $\mathbf{p}$ as the *linear momentum*.

We next write $\mathbf{l}$ in terms of our unit vectors $\hat{\mathbf{a}}$ and $\hat{\mathbf{b}}$.

$$\mathbf{l} = \mathbf{r} \times \mathbf{p} \tag{3.15}$$

$$= \mathbf{r} \times m\dot{\mathbf{r}}$$

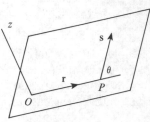

Fig. 3.9

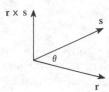

Fig. 3.10

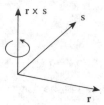

Fig. 3.11

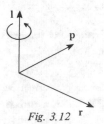

Fig. 3.12

$$= r\hat{\mathbf{a}} \times m(\dot{r}\hat{\mathbf{a}} + r\dot{\theta}\hat{\mathbf{b}})$$

$$= mr^2\,\dot{\theta}\hat{\mathbf{a}} \times \hat{\mathbf{b}}, \qquad \text{since} \qquad \hat{\mathbf{a}} \times \hat{\mathbf{a}} = 0,$$

$$= mr^2\,\dot{\theta}\hat{\mathbf{c}}, \tag{3.16}$$

where $\hat{\mathbf{c}}$ is a unit vector, which with $\hat{\mathbf{a}}$ and $\hat{\mathbf{b}}$ forms a right-handed set of mutually perpendicular unit vectors. This gives the magnitude and direction of l. Differentiating (3.15) with respect to time, we have

$$\dot{\mathbf{l}} = \dot{\mathbf{r}} \times m\dot{\mathbf{r}} + \mathbf{r} \times m\ddot{\mathbf{r}},$$

that is,

$$\dot{\mathbf{l}} = \mathbf{r} \times \mathbf{F}. \tag{3.17}$$

Just as $\mathbf{r} \times \mathbf{p}$ gave the turning effect of $\mathbf{p}$ about O, so $\mathbf{r} \times \mathbf{F}$ gives the turning effect of the force $\mathbf{F}$ about O. It is called the *torque* of $\mathbf{F}$ about O.

Progress test

1. Which one of the following is not a correct expression for the angular momentum about the origin?

   A $\mathbf{l} = \mathbf{r} \times m\dot{\mathbf{r}}$

   B $\mathbf{l} = m\mathbf{r} \times \dot{\mathbf{r}}$

   C $\mathbf{l} = \mathbf{p} \times \mathbf{r}$

   D $\mathbf{l} = mr^2\,\dot{\theta}\hat{\mathbf{a}} \times \hat{\mathbf{b}}.$

2. If $\hat{\mathbf{a}}$ points East and $\hat{\mathbf{b}}$ points South, then $\hat{\mathbf{a}} \times \hat{\mathbf{b}}$ points

   A   North
   B   South
   C   up
   D   down.

3.4   Conservation of angular momentum and energy

In this section, we shall return to the equations of motion (3.12) and treat them from the point of view of the conservation laws.

For a central force $\mathbf{F} = \hat{\mathbf{a}}\,F(r)$, we at once obtain from (3.17),

$$\dot{\mathbf{l}} = \hat{\mathbf{a}}r \times \hat{\mathbf{a}}F(r) = 0,$$

that is,

$$\mathbf{l} = \text{constant}. \tag{3.18}$$

Fig. 3.13

Solutions

1. (B) The force is along the radius vector, i.e., along â, proportional to r and directed backwards.

2. (A) As the force is always along r and proportional to r, it can be expressed as the negative derivative of a potential.

3. (D) It is attractive.

58

[3.4

Thus in motion under a central force, angular momentum about the centre of force is conserved. This can also be seen intuitively, since a central force does not have a torque about the force centre. Returning to (3.16), we see that

$\hat{\mathbf{c}}$ is a constant unit vector, (3.19)

$l = mr^2\dot\theta$ is constant. (3.20)

The first of these shows that the plane perpendicular to $\hat{\mathbf{c}}$, i.e., the plane defined by $\hat{\mathbf{a}}$ and $\hat{\mathbf{b}}$, has a constant direction in space. Hence motion under central forces takes place in a fixed plane. This very important result justifies our use of two-dimensional polar co-ordinates to describe motion under central forces, as was stated in section 3.1.

Equation (3.20) is in fact identical with (3.12b), so that we see that one of our equations of motion is the result of our new conservation law of angular momentum. The other, (3.12a), as we shall now show, follows from the conservation law of energy. This is given by

$T + V = E,$ (3.21)

where E is the total energy, V is the potential energy, defined in (3.10), and

$T = \tfrac{1}{2}mv^2 = \tfrac{1}{2}m(\dot r^2 + r^2\dot\theta^2)$ (3.22)

is the kinetic energy. Hence we have

$\tfrac{1}{2}m(\dot r^2 + r^2\dot\theta^2) + V(r) = E.$ (3.23)

We differentiate with respect to r:

$$m\left(\dot r\frac{\mathrm{d}\dot r}{\mathrm{d}r} + r\dot\theta^2 + r^2\dot\theta\frac{\mathrm{d}\dot\theta}{\mathrm{d}r}\right) + \frac{\mathrm{d}V}{\mathrm{d}r} = 0$$ (3.24)

and eliminate $\mathrm{d}\dot\theta/\mathrm{d}r$ by differentiating (3.20) with respect to r:

$$2r\dot\theta + r^2\frac{\mathrm{d}\dot\theta}{\mathrm{d}r} = 0.$$ (3.25)

Hence, and remembering that

$$\dot r\frac{\mathrm{d}\dot r}{\mathrm{d}r} = \ddot r,$$ (3.26)

we have

$m(\ddot r - r\dot\theta^2) - F(r) = 0,$ (3.27)

which is the other equation of motion (3.12a). We have also made use here of (3.10), i.e., that $F(r)$ is a conservative force, derived from the potential-energy function.

We have thus shown that the two equations of motion are equivalent to the two conservation laws of angular momentum and energy, as applied to the motion. Linear momentum is of course not conserved in a central force motion, since the particle is moving under an external force. The great advantage that lies in the use of the conservation laws is that they are first-order differential equations, while the equations of motion are second-order differential equations. In solving a problem by means of the conservation laws, we therefore save ourselves one integration. On the other hand, in the formulation of the problem they are less general, since they contain the constant values of the angular momentum and energy, which are known as the *constants of motion*. In starting from the equations of motion, these constants arise as constants of integration.

Worked example 3.1. A particle moves under a central force in a circular orbit through the centre of force. Find the law of force.

The equation of the circle is

$$r = 2a \cos \theta \qquad (1)$$

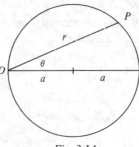

Fig. 3.14

The idea now is to eliminate θ between (1) and the angular momentum equation $r^2 \dot{\theta} = C$. We have

$$\dot{r} = -2a \sin \theta \, \dot{\theta} = -\frac{2aC}{r^2} \sin \theta$$

$$\ddot{r} = \frac{4aC}{r^3} \sin \theta \, \dot{r} - \frac{2aC}{r^2} \cos \theta \, \dot{\theta}$$

$$= -\frac{8a^2 C^2}{r^5} \sin^2 \theta - \frac{2aC^2}{r^4} \cos \theta$$

$$= -\frac{8a^2 C^2}{r^5} \left(1 - \frac{r^2}{4a^2} \right) - \frac{2aC^2}{r^4} \frac{r}{2a}, \text{ using (1)}$$

$$= -\frac{8a^2 C^2}{r^5} + \frac{C^2}{r^3}$$

$$\therefore \quad F = mf = m(\ddot{r} - r\dot{\theta}^2)$$

$$= m\left(-\frac{8a^2 C^2}{r^5} + \frac{C^2}{r^3} - \frac{C^2}{r^3}\right)$$

$$= -\frac{8a^2 C^2 m}{r^5}.$$

The force is thus attractive and inversely proportional to the fifth power of the distance. (The author apologizes for introducing here a highly artificial example, but the motion of particles under physically realistic forces generally requires a computer for its solution.)

Progress test

1. A central force always conserves

   A   energy
   B   linear momentum
   C   angular momentum
   D   none of the above.

3.5   The radial equation

Finally, before turning to physical applications of the theory. we re-write the equations of motion in another form, Eliminating θ between the conservation equations (3.20) and (3.23) we have

$$\tfrac{1}{2}m\dot{r}^2 + \frac{l^2}{2mr^2} + V(r) = E. \tag{3.28}$$

Mathematically, this is equivalent to the motion in one dimension of a particle of mass m and total energy E under a force given by the effective potential energy

$$U(r) = V(r) + \frac{l^2}{2mr^2}, \tag{3.29}$$

except that the equation makes physical sense only for $r \geqslant 0$, since there is no meaning attached to negative values of r. In physical terms, (3.28) describes the motion of a particle under a central force, as seen by an observer at O, who is himself turning so as always to look towards the particle. Such an observer is in a rotating frame of reference, which is clearly not an inertial frame of reference, and the additional potential energy term in (3.29) in fact leads to an inertial force

$$F_c = -\frac{d}{dr}\left(\frac{l^2}{2mr^2}\right) = \frac{l^2}{mr^3}. \tag{3.30}$$

With the help of (3.20) we can eliminate l and put the expression into a more familiar form:

$$F_c = mr\dot{\theta}^2. \tag{3.31}$$

This is the *centrifugal force*, which, as we shall see in chapter 6, always enters when we treat a motion from the point of view of a rotating frame of reference. Note that while the centrifugal force is a central force, it depends on the angular velocity of the frame of reference relative to an inertial frame. Only in the special case, when the real force is central, so that the angular momentum is constant, can we eliminate the angular velocity and derive the centrifugal force from a potential energy.

Since in (3.28) we must have $\dot{r}^2 \geqslant 0$, the possible values of r for given E and l are determined by the inequality

$$U(r) = V(r) + \frac{l^2}{2mr^2} \leqslant E \tag{3.32}$$

with the maximum and minimum values of r, for which $\dot{r} = 0$, given by the equality sign in the above equation. When $U(r)$ has a minimum, and is equal to this minimal value, then $\dot{r} = 0$ throughout the motion, which is therefore circular. For values of E less than this, no motion is possible.

To illustrate the above points, let us consider motion under an attractive inverse square law force. Examples of this are planetary motion or the motion of the electron in the hydrogen atom. We then have, assuming $V(\infty) = 0$,

$$F(r) = -\frac{k}{r^2}, \qquad V(r) = -\frac{k}{r}, \qquad k > 0. \tag{3.33}$$

Hence

$$U(r) = -\frac{k}{r} + \frac{l^2}{2mr^2}. \tag{3.34}$$

To sketch $U(r)$ as a function of r, all that it is necessary to note is that, for large r, the first term is dominant, so that

$$U(r) < 0 \quad \text{for large } r \text{ and} \quad U(r) \to 0 \quad \text{as } r \to \infty,$$

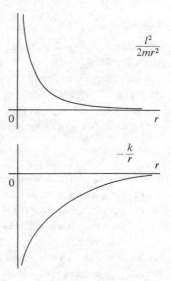

Fig. 3.15

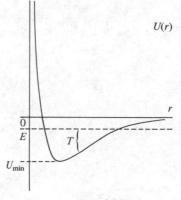

Fig. 3.16

while, for small r, the second term is dominant, so that

$$U(r) \to +\infty \qquad \text{as} \qquad r \to 0.$$

Clearly then, $U(r)$ must be of the form shown. The same graph can also be used to show the constant value E of the energy for a given motion. The kinetic energy T at any distance r can then be read off at once, as can the values of r for which the motion is possible.

We now consider motions for different values of E. In order to show these more clearly, we plot $U(r)$ both to the left and to the right, i.e., for $\theta = 0°$ and $\theta = 180°$. We then have four distinct cases:

(a) $E > 0$. There is a minimum radial distance, but no maximum. For $\theta \to 0$, the path goes off to infinity as shown.

(b) $0 > E > U_{\min}$. There is now both a maximum and a minimum radial distance, so that a possible shape is as shown.

(c) $E = U_{\min}$. This is the limiting case when, as has already been shown, the path is a circle.

(d) $U_{\min} > E$. No motion is possible.

All this can be shown from energy considerations, without solving the equations of motion. When we actually do this, we shall see that (a) gives a hyperbola and (b) an ellipse, with the focus at the centre of attraction. It is also clear that (a) is a free state, in which case we refer to the path as a *trajectory*, while (b) and (c) are bound states and the paths are called *orbits*.

A word of caution must be added regarding diagram (b). It is only in exceptional cases, of which the inverse square law force is one, that the angular distance between successive maxima of the radius vector is exactly $360°$, so that the path retraces itself. More generally, the path has the shape of a rosette. [See Fig. 3.18.]

Worked example 3.2. Show that for the inverse square law of attraction

$$F(r) = -\frac{GMm}{r^2}$$

the orbit with given angular momentum l and least energy E_0 is a circle of radius

$$r_0 = \frac{l^2}{GMm^2}.$$

Solution

1. (A) It conserves angular momentum only about *the centre of force.*

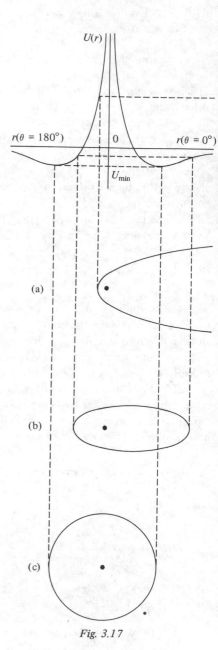

Fig. 3.17

We have

$$U(r) = -\frac{GMm}{r} + \frac{l^2}{2mr^2}$$

$$\therefore \quad \frac{dU}{dr} = \frac{GMm}{r^2} - \frac{l^2}{mr^3}.$$

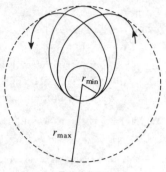

Fig. 3.18

The orbit with least energy is given by

$$E_0 = U_{min}$$

and is a circle. Hence its radius r_0 is given by

$$\frac{dU}{dr} = 0 \quad \text{when} \quad r = r_0,$$

that is,

$$r_0 = \frac{l^2}{GMm^2}.$$

The minimum energy is

$$E_0 = -\frac{GMm}{r_0} + \frac{GMm^2 r_0}{2mr_0^2} = -\frac{GMm}{2r_0} = -\frac{G^2 M^2 m^3}{2l^2}.$$

Progress test

1. If we eliminate r and its derivatives from (3.20) and (3.28), the resulting equation is independent of

   A l
   B m
   C E
   D V.

2. The centrifugal force arises whenever we deal with motion

   A   in a circle
   B   under central forces
   C   under inertial forces
   D   referred to a rotating frame of reference.

3. In motion under a central force we always have circular motion when

    A   the total energy is least
    B   the potential energy is least
    C   the effective potential energy is least
    D   the angular momentum is least.

4. Which one of the following is not a bound state problem?

    A   The motion of Halley's comet
    B   The motion of a moon probe
    C   The motion of the electron in a hydrogen atom
    D   The motion of the α-particle in Rutherford scattering.

3.6   Gravitational attraction

Kepler (1571–1630), by trial and error over a period of many years, established three laws of planetary motion that fitted the enormous amount of data obtained by Tycho Brahe (1546–1601) through pain-staking and accurate observation. These were:

K1.   Each planet moves in an ellipse with the sun as a focus.

K2.   For each planet, the line from the sun to the planet sweeps out equal areas in equal times.

K3.   The squares of the periods of revolution of the planets are proportional to the cubes of the major axes of the ellipses.

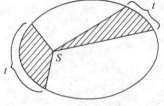

Fig. 3.19

The analysis of these empirical laws by Newton, which eventually led him to his enunciation of the law of universal gravitation, has always been considered one of the truly monumental achievements of the human mind.

We first consider K2. The element of area dA swept out by the radius vector in time dt is approximately $\frac{1}{2}r(r + dr)\,d\theta$. Hence in the limit,

$$dA = \tfrac{1}{2}r^2\,d\theta \quad \text{or} \quad \dot{A} = \tfrac{1}{2}r^2\,\dot{\theta}. \tag{3.35}$$

Now K2 can be re-worded to read that the rate of sweeping out the area A is constant, that is,

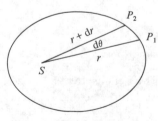

Fig. 3.20

$$r^2\,\dot{\theta} = C \quad \text{or} \quad \frac{d}{dt}(r^2\,\dot{\theta}) = 0. \tag{3.36}$$

Comparing this with (3.12b) we see that K2 is equivalent to stating that the planets move under a central force directed towards the sun.

It will be noticed that we have assumed here that the planets and the sun can be represented by point masses. We shall show in the next section that this is in fact permissible.

We next consider K1. The equation of an ellipse in polar co-ordinates with one focus as origin is

$$1 - \epsilon \cos \theta = \frac{a(1 - \epsilon^2)}{r},$$ (3.37)

where a is the semi-major axis and ϵ the eccentricity of the ellipse. [Experts in conic sections will recognize the quantity $a(1 - \epsilon^2)$ as the *semi-latus rectum*.] We differentiate twice with respect to time and use (3.36) to eliminate $\dot{\theta}$:

$$\epsilon \sin \theta \, \dot{\theta} = -\frac{a(1 - \epsilon^2)}{r^2} \dot{r}.$$

$$\therefore \quad C\epsilon \sin \theta = -a(1 - \epsilon^2) \dot{r}.$$

$$\therefore \quad C\epsilon \cos \theta \, \dot{\theta} = -a(1 - \epsilon^2) \ddot{r}.$$

$$\therefore \quad \frac{C^2}{r^2} \epsilon \cos \theta = -a(1 - \epsilon^2) \ddot{r}.$$ (3.38)

This we now substitute in (3.12a) to obtain, for the central force on a planet of mass m,

$$F(r) = m(\ddot{r} - r\dot{\theta}^2)$$

$$= -\frac{mC^2}{a(1 - \epsilon^2)r^2} \left[\epsilon \cos \theta + \frac{a(1 - \epsilon^2)}{r} \right].$$

Hence, using (3.37), we have

$$F(r) = -\frac{mC^2}{a(1 - \epsilon^2)r^2}.$$ (3.39)

Thus the force is attractive, because of the minus sign, and inversely proportional to the square of the distance.

Finally we relate C to the period T of the planet. We see from (3.35) that

$$\dot{A} = \tfrac{1}{2}C,$$

so that the relation between the area of the ellipse, which can be shown to be

$$A = \pi a^2 (1 - \epsilon^2)^{1/2},$$

and the period T of the planet is

$$\pi a^2 (1 - \epsilon^2)^{1/2} = \tfrac{1}{2}CT.$$ (3.40)

Hence, from (3.39),

$$F(r) = -\frac{4\pi^2\, a^3\, m}{T^2\, r^2} = -\frac{Bm}{r^2},$$

(3.41)

where, according to K3, the quantity

$$B = \frac{4\pi^2\, a^3}{T^2}$$

(3.42)

is the same for all planets.

Newton's great contribution was the realization that the force between the sun and the planets, that between the earth and the moon, and that on an object near the surface of the earth, were of the same kind. Thus if we determine the constant B from the radius a and period T of the moon's orbit round the earth, which is so nearly circular that we can speak of the radius rather than of the semi-major axis, then the acceleration due to gravity at the surface of the earth, radius R, is

$$g = \frac{B}{R^2}.$$

(3.43)

In this formula, all quantities can be measured and excellent agreement is obtained.

Next, the force on any planet, of mass m, due to the sun is

$$F = -\frac{Bm}{r^2}.$$

(3.44)

Since the law of gravitation is to apply equally to the attraction between all bodies, the corresponding force on the sun, of mass M, must be

$$F = -\frac{bM}{r^2}$$

where b depends on the planet, but not on the sun. Hence

$$\frac{B}{M} = \frac{b}{m} = G, \text{ say,}$$

(3.45)

where G is a universal constant. Then

$$F = -G\,\frac{Mm}{r^2}.$$

(3.46)

This is Newton's law of universal gravitation, applicable to the gravitational interaction between any two bodies of masses M and m

Solutions

1. (B) The resulting equation is
$$\tfrac{1}{2}l(\tfrac{1}{4}\dot\theta^2 + \dot\theta^4) + \theta\,3V = \theta\,3E.$$

2. (D) A centrifugal force is an inertial force and hence arises only if a particular non-inertial frame, i.e., a rotating frame, is used.

3. (A) This is bookwork.

4. (D) In an unbound state, the separation between particles becomes infinite. Note that the moon probe is bound in the system that includes both earth and moon.

respectively, distance r apart. The corresponding gravitational potential energy is

$$V = -G\frac{Mm}{r}. \tag{3.47}$$

The currently accepted value of G is

$$G = (6 \cdot 670 \pm 0 \cdot 015) \times 10^{-11}\,\text{Nm}^2\,\text{kg}^{-2}.$$

Combining (3.43) and (3.45), we have

$$GM_E = gR_E^2, \tag{3.48}$$

where the suffix E refers to the earth.

We can now calculate the escape velocity for a satellite that is launched from the surface of the earth or, more accurately, from a point at the top of the atmosphere. For escape to be possible we must have the total energy

$$E = \tfrac{1}{2}mv^2 - \frac{GmM_E}{R_E} \geqslant 0 \tag{3.49}$$

where m and v are the mass and launching velocity respectively of the satellite. Using (3.48), we have

$$v_{\text{escape}} = (2gR_E)^{1/2}, \tag{3.50}$$

which leads to a value of $11 \cdot 2\,\text{km s}^{-1}$ for the escape velocity.

Progress test

1. The force of gravitation between two point masses m and M a distance r apart is

    A GmM/r^2

    B $-GmM/r^2$

    C GmM/r

    D $-GmM/r$.

2. If a particle of mass m moves in a circle of radius b under an inverse square attraction $F = -k/r^2$, then the velocity of the particle is

    A k/mb

    B $\sqrt{k/mb}$

    C $\sqrt{k/b}$

    D   none of these.

3. The total energy at any point of the orbit of the particle in question 2 is

A k/b

B $-k/b$

C $-k/2b$

D   none of these.

4. The escape velocity for a satellite is

A   greatest for a horizontal launching
B   greatest for a vertical launching
C   least for a vertical launching
D   none of these.

3.7   Gravitational force due to a sphere

We shall now prove the result used in the last section that the gravitational attraction due to a uniform sphere at a point outside the sphere is equal to that due to a point particle of the same mass at the centre of the sphere.

We begin by considering the gravitational potential energy of a particle due to a spherical shell of mass M_s, radius R_s; the particle, of mass m, is distance r from the centre of the shell. The potential energy is obtained by summing the potential energies due to ring elements, like the one shown shaded. The area of such an element is $2\pi R_s^2 \sin \theta \, d\theta$ and hence its mass is

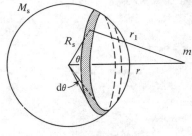

Fig. 3.21

$$dM_s = \frac{2\pi R_s^2 \sin \theta \, d\theta}{4\pi R_s^2} M_s = \tfrac{1}{2} M_s \sin \theta \, d\theta.$$

The potential energy due to this element is

$$dV_s = -G \frac{m(\tfrac{1}{2} M_s \sin \theta \, d\theta)}{r_1}. \qquad (3.51)$$

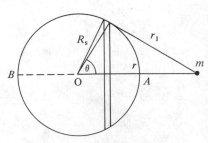

Fig. 3.22

Now

$$r_1^2 = R_s^2 + r^2 - 2R_s r \cos \theta,$$

so that

$$r_1 \, dr_1 = R_s r \sin \theta \, d\theta.$$

When this is substituted in (3.51) we have

$$dV_s = -\frac{GmM_s\, dr_1}{2R_s r}.$$

This must be integrated over all r_1 from A to B. We distinguish two cases:

(a)   The mass m is outside the shell, $r > R_s$. Then

$$V_s(r) = -\frac{GmM_s}{2R_s r}\int_{r-R_s}^{r+R_s} dr_1 = -\frac{GmM_s}{r}, \quad r > R_s. \qquad (3.52)$$

(b)   The mass is inside the shell, $r < R_s$. Then

$$V_s(r) = -\frac{GmM_s}{2R_s r}\int_{R_s-r}^{R_s+r} dr_1 = -\frac{GmM_s}{R_s}, \quad r < R_s. \qquad (3.53)$$

The force on the mass m is

$$F_s(r) = -\frac{dV_s}{dr} = \begin{cases} -\dfrac{GmM_s}{r^2}, & r > R_s, \\[2mm] 0, & r < R_s. \end{cases} \qquad (3.54)$$

Thus the force due to a shell at a point outside the shell is the same as that due to a particle of equal mass at the centre of the shell, and there is no force inside the shell.

    Turning now to the potential energy due to a sphere, mass M and radius R, we can build up the sphere from concentric shells of volume $4\pi R_s^2\, dR_s$. Hence the mass of each shell is

$$dM = \frac{4\pi R_s^2\, dR_s}{\frac{4}{3}\pi R^3} M = \frac{3R_s^2\, M}{R^3}\, dR_s.$$

We again distinguish two cases:

(a)   The mass M is outside the sphere, $r > R$. Then

$$V(r) = -\int_0^R \frac{Gm}{r}\frac{3R_s^2\, M}{R^3}\, dR_s = -\frac{GmM}{r}, \qquad r > R. \qquad (3.55)$$

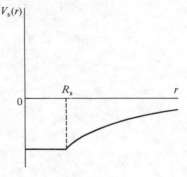

Fig. 3.23

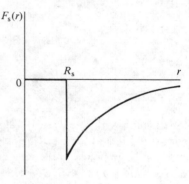

Fig. 3.24

70

(b) The mass m is inside the sphere, $r < R$. Then

$$V(r) = -\int_0^r \frac{Gm}{r}\frac{3R_s^2 M}{R^3}\,dR_s - \int_r^R \frac{Gm}{R_s}\frac{3R_s^2 M}{R^3}\,dR_s.$$

$$\therefore \quad V(r) = -\frac{GmM}{2R^3}(3R^2 - r^2), \qquad r < R. \tag{3.56}$$

It should be noted that we have assumed here that the law of gravitation, which has been verified only for mass points in empty or near empty space, also holds for mass points embedded in dense matter.

As before, we obtain the force on the particle,

$$F(r) = \begin{cases} -\dfrac{GmM}{r^2}, & r > R, \\[2ex] -\dfrac{GmMr}{R^3}, & r < R. \end{cases} \tag{3.57}$$

The explanation of the negative sign in both (3.54) and (3.57) is that the force is directed towards O. Force is of course a vector quantity, and the negative sign indicates the direction of this vector. The results (3.55) and (3.57) are the ones which enable us to treat spheres as mass points from the point of view of their gravitational attraction at points outside themselves. Although Newton suspected this result to be true as early as 1665, he could not find a satisfactory proof for many years. For that reason he apparently refused to publish any of his work on gravitation until 1683. Such patience is given to few scientists.

Worked example 3.3. A satellite of mass m, initially in a circular parking orbit, radius r, accelerates along its line of motion to change to another circular orbit, radius $s > r$. Describe the procedure.

The rocket motor is first run at A in such a way that the satellite goes into an elliptic orbit. Since at that point, the orbit is perpendicular to the radius vector, A is the point on the orbit nearest to the origin. When it reaches the point B, furthest from the origin it is again moving at right angles to the radius vector. The motor is now run again to bring the velocity to the necessary value for circular motion with radius s.

Let the velocities at A before and after the motor is run be u, v; and at B let them be w, x. Let the mass of the earth be M. Then for circular motion,

$$\frac{mv^2}{r} = \frac{GMm}{r^2} \quad \text{and} \quad \frac{mx^2}{s} = \frac{GMm}{s^2}. \tag{1}$$

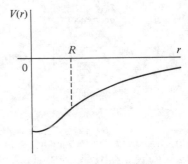

Fig. 3.25

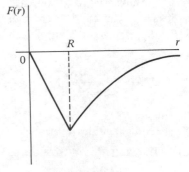

Fig. 3.26

For the elliptic motion we have conservation of angular momentum

$$mvr = mws \qquad (2)$$

and of energy

$$\tfrac{1}{2}mv^2 - \frac{GMm}{r} = \tfrac{1}{2}mw^2 - \frac{GMm}{s}. \qquad (3)$$

(We assume that the satellite loses negligible mass through the running of the motor and the change in the motion is instantaneous.) From (2) and (3) we obtain

$$v^2 = \frac{2GMs}{r(s+r)}, \qquad w^2 = \frac{2GMr}{s(s+r)}.$$

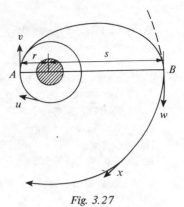

Fig. 3.27

Hence the gain in energy at A is

$$\tfrac{1}{2}m(v^2 - u^2) = \frac{GMm}{2r}\frac{s-r}{s+r},$$

and at B is

$$\tfrac{1}{2}m(x^2 - w^2) = \frac{GMm}{2s}\frac{s-r}{s+r}.$$

Hence the total gain in energy is

$$\Delta E = \frac{GMm(s-r)}{2rs} = \frac{GMm}{2}\left(\frac{1}{r} - \frac{1}{s}\right).$$

This last result could have been obtained at once by considering the change in total energy in going from one circular orbit to another:

$$\Delta E = \left(\tfrac{1}{2}mx^2 - \frac{GMm}{s}\right) - \left(\tfrac{1}{2}mu^2 - \frac{GMm}{r}\right)$$

$$= -\frac{GMm}{2s} + \frac{GMm}{2r}.$$

Note that at no point in the argument did we need to know that AB was part of an ellipse. All that was needed were the conservation laws.

Progress test

1. The gravitational potential energy of a particle inside a spherical shell
    A   is greater at the centre
    B   is least at the centre
    C   is constant, but non-zero, throughout the volume of the shell
    D   is zero throughout the volume of the shell.

2. The gravitational force on a particle inside a sphere of uniform density is

    A   proportional to the distance from the centre and directed towards it
    B   proportional to the distance from the centre and directed away from it
    C   inversely proportional to the distance from the centre
    D   none of these.

3.8   The gravitational field

In the last section we calculated the gravitational force on a particle due to a sphere. Quite generally, the gravitational force on a particle due to a distribution of matter depends on both the particle and the matter distribution, and it would be useful to separate out the effects of the two. We therefore define the *field* due to a matter distribution as the force on a particle of unit mass due to the distribution, and we similarly define the *potential* due to a matter distribution as the potential energy of a particle of unit mass due to the distribution. Clearly we obtain the formulae for the field and potential due to a sphere simply by putting $m = 1$ in (3.55)–(3.57). An entirely analogous use of the concept of field occurs in electricity, where it refers to the force on unit charge.

The value of the concepts of potential and field is that they depend only on the matter distribution which causes them. Their effect is on a second matter distribution, e.g., we speak of the earth moving in the gravitational field of the sun. Thus we separate the cause—potential or field due to the sun—from the effect—motion of the earth, while as long as we consider the situation in terms of forces, this is not possible, since the force between two bodies, e.g. sun and earth, depends on them both. The potential due to a matter distribution is a scalar function of position, $\phi(\mathbf{r})$, by which we mean that at every point of space there is specified a scalar quantity, ϕ, which varies from point to point and hence is a function of the vector $\mathbf{r}$, and which we call the potential. Similarly the field is a vector function of position, $\mathbf{g}(\mathbf{r})$, i.e., at every point of space there is specified a vector, $\mathbf{g}$, which varies in both magnitude and direction from point to point. Often, for complicated matter distributions, we may specify the potential and field directly rather than the matter distribution which has caused it. A simple example is the constant gravitational field of magnitude g near the surface of the earth, which is specified without any direct reference to the matter of the earth which causes it.

We originally defined potential energy as the negative of work, which in turn was obtained from the product of force and distance. We return to this definition in order to establish a relationship between potential and field. If the field vector g moves through a distance ds along a given path, then the change in potential is given by

$$d\phi = -\mathbf{g} \cdot \mathbf{ds}. \tag{3.58}$$

The potential at a point P can then be obtained, starting from the potential at an arbitrary point A, by summing the small quantities $d\phi$ and proceeding to the limit,

$$\phi_P = -\lim_{ds \to 0} \sum_{A}^{P} \mathbf{g} \cdot \mathbf{ds} = -\int_{A}^{P} \mathbf{g} \cdot \mathbf{ds}, \tag{3.59}$$

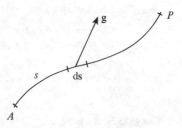

Fig. 3.28

where the integral is taken along the path AP. Now let us take two different paths from A to P and label them (1) and (2). Then the gravitational field is such that the quantity ϕ_P depends only on the positions of O and P and not on the particular path of integration, that is

$$\int_{A(1)}^{P} \mathbf{g} \cdot \mathbf{ds} = \int_{A(2)}^{P} \mathbf{g} \cdot \mathbf{ds}. \tag{3.60}$$

This is the general definition of a conservative field. It follows at once that there is no change in ϕ, when $\mathbf{g} \cdot \mathbf{ds}$ is taken round a closed loop, since

$$\left(\int_{A(1)}^{P} + \int_{P}^{A(2)} \right) \mathbf{g} \cdot \mathbf{ds} = 0. \tag{3.61}$$

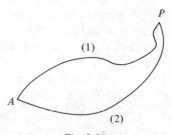

Fig. 3.29

This means that no work is done in taking a particle round a closed loop in a gravitational field, so that the gravitational field conserves energy.

Our definition of a conservative field is an extension of our earlier definitions for forces in one dimension and for central forces. For a central field, (3.58) can be written

$$d\phi = -g\hat{\mathbf{a}} \cdot \mathbf{ds} = -g\, dr \tag{3.62}$$

where $\hat{\mathbf{a}}$ is as before the unit vector along the radius vector $\mathbf{r}$, and g is a function of r only. Hence for elements ds_1 and ds_2, equidistant from the centre O of the field, of two paths $A(1)P$ and $A(2)P$,

$$\mathbf{g}_1 \cdot \mathbf{ds}_1 = g(r)\, dr,$$

$$\mathbf{g}_2 \cdot \mathbf{ds}_2 = g(r)\, dr,$$

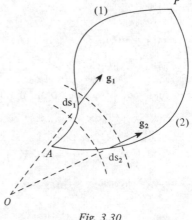

Fig. 3.30

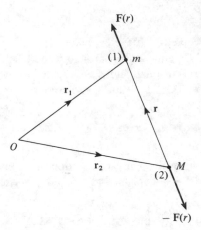
Fig. 3.31

74

and this is true for all elements of the paths. Hence any central field is conservative and

$$g(r) = -\frac{d\phi}{dr}, \tag{3.63}$$

which is equivalent to (3.10).

Solutions

1. (C) It is equal to the value on the surface of the shell.

2. (A) The force is attractive and increases linearly with distance.

Progress test

1. Force as a function of position is

    A   a vector function of a scalar
    B   a vector function of a vector
    C   a scalar function of a vector
    D   a scalar function of a scalar.

3.9   The two-body problem

So far we have assumed that, when a planet moves round the sun or the moon moves round the earth, the heavier body remains at rest in some inertial frame. This is in fact not so, as we shall now show.

Consider, in general, a system of two particles (1) and (2) with masses m and M and position vectors $\mathbf{r}_1$ and $\mathbf{r}_2$ in some inertial frame of reference, and let them move under a mutual force between them. Then the force $\mathbf{F}(r)$ on particle (1) due to particle (2) is a function of the distance r between them, where

$$\mathbf{r} = \mathbf{r}_1 - \mathbf{r}_2, \tag{3.64}$$

and will be along the direction of $\mathbf{r}$. By Newton's third law, the force on particle (2) due to particle (1) is $-\mathbf{F}(r)$. Then, if there are no external forces on the system, the equations of motion of the two particles are

$$m\ddot{\mathbf{r}}_1 = \mathbf{F}(r), \tag{3.65}$$

$$M\ddot{\mathbf{r}}_2 = -\mathbf{F}(r). \tag{3.66}$$

From these equations we obtain the further equations

$$m\ddot{\mathbf{r}}_1 + M\ddot{\mathbf{r}}_2 = 0, \tag{3.67}$$

$$mM(\ddot{\mathbf{r}}_1 - \ddot{\mathbf{r}}_2) = (m + M)\mathbf{F}(r). \tag{3.68}$$

The first of these can be integrated at once and gives

$$m\dot{\mathbf{r}}_1 + M\dot{\mathbf{r}}_2 = \text{constant}. \tag{3.69}$$

In other words, the total linear momentum of the system is constant in the inertial frame that we are using, as would be expected, since there are no external forces acting on the system. On the other hand, $\dot{\mathbf{r}}_2$ is clearly not constant, so that the frame of reference in which particle (2) is at rest is not an inertial frame.

Let us now choose that inertial frame in which the constant in (3.69) is zero. Then, integrating again, we obtain

$$m\mathbf{r}_1 + M\mathbf{r}_2 = \text{constant}. \tag{3.70}$$

Hence in this frame we have a constant position vector, which we define through

$$\mathbf{R} = \frac{m\mathbf{r}_1 + M\mathbf{r}_2}{m + M}. \tag{3.71}$$

This is the position vector of a point C, which is called the *centre of mass* of the system. [See Fig. 3.33.] Thus the frame we have chosen is the one in which the centre of mass of the system is at rest. It is always referred to as the *CM-frame*. It may be noted that it is an inertial frame only if there are no external forces on the system.

We now turn to (3.68). Using (3.64), it can be written

$$\mu\ddot{\mathbf{r}} = \mathbf{F}(r), \tag{3.72}$$

where

$$\mu = \frac{mM}{m + M} \tag{3.73}$$

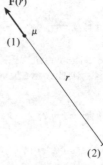

F(r)

μ

(1)

r

(2)

Fig. 3.32

is called the *reduced mass* of the system, since it is always less than either m or M. We see that (3.72) describes the motion of a particle of mass μ at position (1) under a central force $\mathbf{F}(r)$, with the centre of force and origin of the reference frame at (2). As far as the mathematical formalism goes, this is exactly the state of affairs dealt with in section 3.2. Thus we can still describe the motion in terms of the equations (3.12a) and (3.12b), provided that we replace the mass m by the reduced mass μ on the right-hand side of (3.12a). A word of caution is appropriate here. The replacement of m by μ results from the kinematics of the situation, i.e., it is a question of how we describe the motion of the particle. It has nothing to do with the dynamics, i.e., what causes the motion. Thus if the force on the particle depends on its mass, then this will still be m and not μ. This is, of course, the case with the gravitational interaction

$$\mathbf{F}(r) = -G\frac{mM}{r^2}\hat{\mathbf{a}}. \tag{3.74}$$

Hence the motion of the fictitious particle with reduced mass is under the same force as the true motion and (3.41) now becomes

vectors.

1. (B) Both force and position are

Solution

$$F(r) = -G\frac{mM}{r^2} = -\frac{4\pi^2 a^3 \mu}{T^2 r^2}, \qquad (3.75)$$

so that

$$\frac{T^2}{a^3} = \frac{4\pi^2}{G(m + M)}, \qquad (3.76)$$

where a is the semi-major axis of the orbit of (1) relative to (2). This leads to a correction to the quantity B in (3.45), i.e., we now have

$$B = \frac{4\pi^2 T^2}{a^3} = G(m + M), \qquad (3.77)$$

so that B is no longer the same for all planets. Kepler's third law is therefore only approximately true, although the approximation is extremely good, since the mass of even the largest planet is very much less than that of the sun. More pronounced departures from Kepler's third law are observed in the corresponding problem of the motion of a planet and its moons and this has made it possible to calculate the masses of certain planets.

When eq. (3.72) has been solved, it is always possible to obtain the motion in terms of the original co-ordinates $\mathbf{r}_1$ and $\mathbf{r}_2$, since it follows from (3.64) and (3.71) that

$$\mathbf{r}_1 = \mathbf{R} + \frac{M}{m + M}\mathbf{r}, \qquad \mathbf{r}_2 = \mathbf{R} - \frac{m}{m + M}\mathbf{r}. \qquad (3.78)$$

If we choose the centre of mass C to be the origin in the *CM-frame*, i.e., if we put $\mathbf{R} = 0$, then we have

$$\mathbf{r}_1 = \frac{M}{m + M}\mathbf{r}, \qquad \mathbf{r}_2 = -\frac{m}{m + M}\mathbf{r}. \qquad (3.79)$$

Hence, as observed from C, particles (1) and (2) perform orbits which are similar to their relative orbit, but reduced in scale by factors $M/(m + m)$ and $m/(m + M)$ respectively. It can then be shown quite easily that, for instance, the earth and moon move in ellipses with C as one focus. Because of the very much larger mass of the earth, C is actually inside the earth, about 4800 km from its centre.

One important application of the above concerns the origin of the tides, for which we shall use the frame of reference centred on C in which the earth and moon are at rest. This is a rotating frame with respect to the fixed stars, and therefore not inertial. The forces on the water in the oceans that cause the tides are now two-fold. Firstly, there

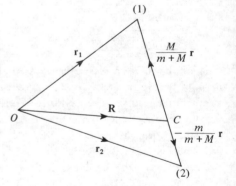

Fig. 3.33

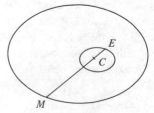

Fig. 3.34

is the attraction of the moon. This attracts the water nearest to the moon more than the rest of the earth and the water furthest from the moon less. The result is that the water nearest to the moon is pulled away from the earth, while the earth is in turn pulled away from the water furthest from the moon. The resulting bulges are reinforced by the centrifugal force (3.31), which has a greater effect on the water furthest from the moon. As the earth rotates once every day on its axis, any point on the earth meets high tide twice a day.

So far, we have assumed that there are no external forces on the earth–moon system. In reality, both move under the attraction from the sun—as well as the much lesser ones from the other planets—and it can then be shown that the centre of mass C moves in an ellipse round the sun. (The mass of the sun is so much greater than that of the earth, that its motion can be neglected.) Viewed from the sun, the earth therefore moves in an ellipse with C as focus, and C in turn moves in an ellipse with the sun as focus.

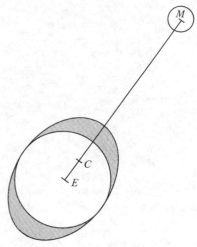

Fig. 3.35

Finally, we evaluate the total angular momentum and energy relative to C. We have, from (3.79), that

$$\mathbf{l} = m\mathbf{r}_1 \times \dot{\mathbf{r}}_1 + M\mathbf{r}_2 \times \dot{\mathbf{r}}_2 = \mu\mathbf{r} \times \dot{\mathbf{r}}, \tag{3.80}$$

$$E = \tfrac{1}{2}m\dot{r}_1^2 + \tfrac{1}{2}M\dot{r}_2^2 + V(r) = \tfrac{1}{2}\mu\dot{r}^2 + V(r). \tag{3.81}$$

Thus the constants of motion of the system in the *CM-frame* are obtained in exactly the same way as in section 3.4, except that we must now use the reduced mass of the system. Applications of this result are common in atomic physics. It leads, for instance, to modifications of the Rydberg constant of hydrogen-like atoms, since the reduced mass of the electron differs for the different nuclear masses in H, He$^+$, Li^{++} etc. [This is discussed in more detail in Jackson, chapter 4.] For positronium, which is a hydrogen-like atom consisting of an electron and a positron, the latter having the same mass as the electron but opposite charge, the reduced mass is half the electron mass, so that the Rydberg constant for positronium is half that for hydrogen. The spectra of atomic hydrogen and of positronium are therefore qualitatively the same, but the spacing between the energy levels are almost exactly twice as large in hydrogen as in positronium.

Worked example 3.4. For small departures from equilibrium, the potential-energy function for the vibration of a diatomic molecule is given by

$$V(r) = \tfrac{1}{2}K(r - r_0)^2, \qquad r - r_0 \ll r_0 *$$

where r is the distance between the two atoms of the molecule and r_0 is the distance at equilibrium. Show that the period of small vibrations about equilibrium is

* The symbol $\ll$ means 'very much less than'.

$$\tau = 2\pi \sqrt{\frac{\mu}{K}}$$

where μ *is the reduced mass of the two atoms.*

The force between the atoms is

$$F(r) = -\frac{dV}{dr} = -K(r - r_0).$$

Hence from (3.72), the equation of motion is

$$\mu\ddot{r} = -K(r - r_0).$$

This is the equation for simple harmonic motion about the point $r = r_0$ and leads to

$$r = A \sin \left(\sqrt{\frac{K}{\mu}} t + \epsilon \right) + r_0,$$

where A and ϵ are constants of integration. Hence the period of vibration is

$$\tau = 2\pi \sqrt{\frac{\mu}{K}}.$$

It may be noted that a molecule is a system that should be treated according to the laws of quantum mechanics and not according to those of classical mechanics. For a harmonic motion the quantum-mechanical treatment leads to the result that the molecule emits photons of energy $h\nu$, where ν is the same frequency as is obtained in the classical calculation, that is,

$$\nu = \frac{1}{2\pi} \sqrt{\frac{K}{\mu}}.$$

Progress test

1. Which one of the following is not correct for the centre of mass of two particles moving under a mutual interaction?

    A   It lies on the line joining the two particles.
    B   It is nearer to the lighter of the two particles
    C   It moves with uniform velocity in all inertial reference frames
    D   It is at rest in the CM frame.

3.10 Problems

3.1. If a communication satellite is to remain in orbit constantly above a particular city on the equator, what distance above this city will this orbit be? Why must the city be on the equator?

3.2. Calculate the escape velocity of a particle at the surface of the earth and the moon, given that the radii of the earth and the moon are 6380 km and 1740 km, and the mass of the earth is 81 times that of the moon.

3.3. A satellite is launched horizontally at a distance d_0 from the centre of the earth, radius $R < d_0$, with velocity v_0. Show that the distance d and velocity v at another point at which it moves at right angles to the line joining it to the centre are given by $d = \alpha^{-1}d_0, v = \alpha v_0$, where $\alpha = (2gR^2/d_0v_0) - 1$. Can there be more than two points at which the satellite moves at right angles to the line joining it to the centre of the earth?

3.4. In problem 3.3, discuss the following cases and sketch the orbit in each of them:

(a) $v_0^2 < \dfrac{gR^2}{d_0}$ (b) $v_0^2 = \dfrac{gR^2}{d_0}$

(c) $\dfrac{gR^2}{d_0} < v_0^2 < \dfrac{2gR^2}{d_0}$ (d) $v_0^2 = \dfrac{2gR^2}{d_0}.$

3.5. The motion of a particle under a certain force in a plane at time t is given in rectangular co-ordinates by

$$x = a\cos t, \qquad y = a\sin t,$$

where a can have any positive value. Show that the particle moves under a central force directed towards the origin, and that the force is proportional to the distance from the origin and directed towards it.

3.6. The forces between two atoms, distance r apart, can be described by the potential energy function

$$V(r) = -\frac{A}{r^6} + \frac{B}{r^{12}},$$

where A and B are positive constants. Sketch the graph of $V(r)$ against r and then answer the following questions:

(a)   In which region is the interatomic force attractive and in which is it repulsive?
(b)   What is the least value of the total energy of a molecule consisting of the two atoms?
(c)   If the two atoms are released from rest an infinite distance apart, what is their subsequent distance of closest approach?

3.7. Show that if the density of the earth, mass M and radius R, at distance r from the centre is given by

particle.

1. (B) It is nearer to the heavier

Solution

$$\rho(r) = \begin{cases} \rho_0(1 - \alpha r), & r < R \\ 0 & r > R \end{cases}$$

where ρ_0 and α are constant, then the gravitational potential at a point inside the earth, distance r from the centre, is

$$V(r) = -\frac{GM}{(4 - 3\alpha R) R^3} [6R^2 - 2r^2 - \alpha(4R^3 - r^3)].$$

3.8. If it is assumed that the earth has constant density, and if a straight tunnel is cut through the earth between any two points on the surface, show that the motion of a particle in this tunnel is simple harmonic.

Hence show that the time taken for a particle to traverse such a tunnel between any two points on the earth's surface is constant and equal to half the time taken by a satellite to orbit the earth once in an orbit close to the surface of the earth. (You might like to consider the implications of this result for an international transport timetable.)

3.9. Show that the linear momenta of two particles referred to their CM-frame are equal and opposite.

3.10. If in the worked example 3.4, the observed vibration frequency for the molecule NaCl is $\nu = 1.14 \times 10^{13}$ s^{-1}, show that the spring constant K is 120 Nm^{-1}. If the molecule is stretched through 0.25 Å, show that the work done is about 0.2 eV. [Atomic weights of Na and Cl are 23.0 and 35.5; mass of H atom = 1.67×10^{-27} kg, 1 Å = 10^{-10} m. 1 eV = 1.60×10^{-19} J.]

*3.11. If $\hat{i}, \hat{j}, \hat{k}$ are a right-handed set of mutually perpendicular unit vectors, along the rectangular co-ordinate axes Ox, Oy, Oz, show that the angular momentum $\mathbf{l} = \mathbf{r} \times \mathbf{p}$ of a particle can be expressed as $\mathbf{l} = (yp_z - zp_y)\hat{i} + (zp_x - xp_z)\hat{j} + (xp_y - yp_x)\hat{k}$.

4. Collisions between particles

4.1 The conservation laws

In the last chapter we indicated some of the situations in which it was advantageous to use the laws of conservation of energy, momentum and angular momentum, rather than the equations of motion. This is particularly true in collision processes, which are the main tool for the study of atomic and sub-atomic particles. In these processes, the force law is frequently not known, so that it is impossible to write down the equation of motion. On the other hand, the energies and momenta of the colliding particles can be measured both before and after the collision, and much information can be obtained from these measurements. The word 'collision' is used here in a very general sense and does not imply actual contact between the particles. What usually happens is that each is deflected from its original path by the mutual interaction between them.

We write the energy of a particle in free space as

$$E = E_i + \frac{p^2}{2m}. \tag{4.1}$$

Here, E_i is the internal energy of the particle (which, for instance, may be an atom which itself consists of sub-atomic particles in different arrangements) and

$$\frac{p^2}{2m} = \tfrac{1}{2}mv^2 \tag{4.2}$$

is its kinetic energy. Except for section 4.5 we shall confine ourselves to non-relativistic energies.

4.2 Disintegration of a particle

The simplest free two-particle process is the breaking up of a single particle into two, either spontaneously as in nuclear fission, or through an internal explosion. The process is best described in the frame of reference in which the original particle is at rest. Then the momenta of the final particles are equal and opposite, $\mathbf{p}_0$ and $-\mathbf{p}_0$. (See problem 3.9.) Hence energy conservation gives

$$E_i = E_{1i} + \frac{p_0^2}{2m_1} + E_{2i} + \frac{p_0^2}{2m_2}, \tag{4.3}$$

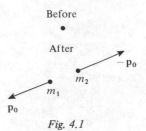

Fig. 4.1

where E_i, E_{1i} and E_{2i} are the internal energies of the initial and two final particles. The *disintegration energy* is defined as

$$\epsilon = E_i - E_{1i} - E_{2i} = \tfrac{1}{2} p_0^2 \left(\frac{1}{m_1} + \frac{1}{m_2} \right)$$

$$= \frac{p_0^2}{2\mu}, \tag{4.4}$$

where

$$\mu = \frac{m_1 m_2}{m_1 + m_2} \tag{4.5}$$

is the reduced mass of the two final particles. [See (3.73).] Clearly $\epsilon > 0$ always.

It may be noticed that we have not included any energy of inter-action between the two final particles in (4.3). The reason for this is that in practice the measurement of the final momenta is made when the particles are so far apart that they no longer interact.

4.3 CM- and LAB-frames

The frame of reference used in the disintegration problem is one in which the centre of mass of the system remained at rest. This is the *CM-frame*, which we have already found useful in section 3.9. In this chapter we shall use the suffix 0 to denote momenta and velocities in it. Another frame of reference, which is particularly useful in collision problems, is one fixed in the laboratory of the experimenter, *the LAB-frame*. In collision experiments, a beam of particles is often produced in a particle accelerator [see Jackson, chapter 8] and these particles then impinge on particles in a target, which is at rest in the laboratory. Thus in many collision problems, the target particle is at rest in the LAB-frame.

Let us return to the problem of the disintegrating particle. Let it have a velocity $\mathbf{V}$ in the LAB-frame before it disintegrates, and let the velocity of one of the particles resulting from the disintegration be v and v_0 in the LAB- and CM-frames respectively. (In this situation the particle is said to disintegrate 'in flight'.) Then

$$\mathbf{v} = \mathbf{V} + \mathbf{v}_0$$

or

$$v^2 + V^2 - 2vV\cos\theta = v_0^2 \qquad (4.6)$$

where θ is the angle between $\mathbf{v}$ and $\mathbf{V}$. This simply expresses the fact that the velocity of the resulting particle in the LAB-frame equals the velocity in the CM-frame plus the velocity of the CM-frame relative to the LAB-frame, which is $\mathbf{V}$. It follows that

$$|V - v_0| \leqslant v \leqslant |V + v_0|. \qquad (4.7)$$

We can also put limits on the kinetic energy of each particle. Since $\mathbf{V}$ is given, and v_0 is uniquely determined by (4.4) and (4.2), the kinetic energy of particle 1 in the LAB-frame is limited by

$$\tfrac{1}{2}m_1(V - v_0)^2 \leqslant T_{LAB} \leqslant \tfrac{1}{2}m_1(V + v_0)^2, \qquad (4.8)$$

and similarly for m_2.

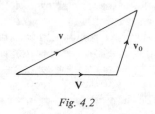

Fig. 4.2

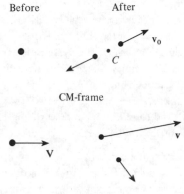

Fig. 4.3

Progress test

1. Which one of the following, in a collision experiment, is moving in both the CM- and LAB-frames?

    A   Incident particle
    B   Target particle
    C   Centre of mass of incident and target particle
    D   The observer.

2. Which of the following diagrams is correct for the motion of two particles relative to their centre of mass C?

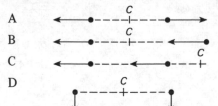

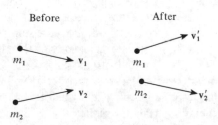

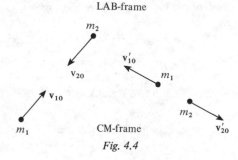

Fig. 4.4

4.4   Elastic collisions

We next treat the problem of the collision of two particles, of masses m_1 and m_2. We shall use suffixes 1 and 2 for the two particles, and unprimed and primed symbols for before and after the collision in the LAB-frame. An additional suffix 0 denotes quantities in the CM-frame.

We shall use the term *collision* to denote not only the interaction between two particles that make a momentary point contact, such as occurs between colliding billiard balls, but also the motion of two particles, such as say an electron and proton, which are deflected from their original paths by a long-range interaction, without ever coming into point contact.

Because the forces on the particles consist simply of the inter-action between them, the momentum for the system of two particles as a whole is conserved in the collision. On the other hand, mechanical energy is not, in general, conserved. Let us consider the system a long time before and after the collision, when the distance between the particles is so large that the forces between them, and hence the potential energy of the system, is zero. (We speak of the particles being 'at infinity' before and after the collision.) The statement that in general mechanical energy is not conserved in a collision means that the total kinetic energy of the particles at infinity after the collision is less than the total kinetic energy of the particles at infinity before the collision. Kinetic energy can be lost in two ways:

(a)   Changes into other forms of energy that are lost to the system, such as heat and sound radiated in the collision of billiard balls or the emission of a photon in an atomic collision;

(b)   Changes in the internal energies of the particles, such as the heating up of billiard balls in a collision, or the exitation of an atom to an excited state in a collision with an electron.

If neither type of change occurs, then we speak of an *elastic collision*. Collisions between macroscopic particles, such as billiard balls, can never be perfectly elastic, because some mechanical energy is always lost, e.g., to the internal energy of the billiard balls, since these heat up. For atomic and sub-atomic particles this is not so. Their energy is quantized [see Jackson, chapter 5] and so a definite amount of energy has to be supplied to raise the internal energy. If the total mechanical energy available is less than that, then the collision can only be elastic. In fact, elastic collisions form one of the major tools for investigating sub-atomic systems, and we shall now consider the theory governing them.

As there are no changes in the internal energies, i.e., as

$$E_1 = E_1', \qquad E_2 = E_2', \tag{4.9}$$

it follows from (4.1) that the total kinetic energy is

$$T = \frac{p_1^2}{2m_1} + \frac{p_2^2}{2m_2} = \frac{p_1'^2}{2m_1} + \frac{p_2'^2}{2m_2} = T', \tag{4.10}$$

and similarly for quantities in the CM-frame. We have already obtained an expression for the position vectors of the particles in this frame in

(3.79), and on differentiating this equation with respect to time, we have

$$\mathbf{v}_{10} = \frac{m_2\,\mathbf{v}}{m_1 + m_2},$$

$$\mathbf{v}_{20} = -\frac{m_1\,\mathbf{v}}{m_1 + m_2}, \tag{4.11}$$

where, following (3.64), we have written

$$\mathbf{v} = \mathbf{v}_{10} - \mathbf{v}_{20} \tag{4.12}$$

as the velocity of particle 1 relative to particle 2. Relative velocities are the same in all frames of reference, since they depend only on successive positions of the particles relative to each other, and not relative to some reference frame. Hence we also have

$$\mathbf{v} = \mathbf{v}_1 - \mathbf{v}_2. \tag{4.13}$$

We again introduce the expression (3.73) for the reduced mass of the system,

$$\mu = \frac{m_1\,m_2}{m_1 + m_2}, \tag{4.14}$$

and then have from (4.11) for the momenta in the CM-frame,

$$\mathbf{p}_{10} = \mu\mathbf{v}, \qquad \mathbf{p}_{20} = -\mu\mathbf{v}, \tag{4.15}$$

so that the total momentum of the system in the CM-frame is zero. Let us denote $\mathbf{p}_{10}$ by $\mathbf{p}_0$, $\mathbf{p}'_{10}$ by $\mathbf{p}'_0$. Then it follows from (4.15) that the lengths of the vectors are equal, that is,

$$p_{10} = p_{20} = p_0, \tag{4.16}$$

and from (4.10) that

$$p_0 = p'_0,$$

$$T_0 = \frac{p_0^2}{2\mu} = \frac{p_0'^2}{2\mu} = T'_0. \tag{4.17}$$

Thus the magnitudes of the momenta, and also the velocities, of the two particles in the CM-frame are unchanged in the collision. The way the collision proceeds in the CM-frame is shown in the figure.

In the LAB-frame, we must add the velocity $\mathbf{V}$ of C to the velocities in the CM-frame. This is obtained by differentiating (3.71):

$$\mathbf{V} = \frac{m_1\,\mathbf{v}_1 + m_2\,\mathbf{v}_2}{m_1 + m_2}. \tag{4.18}$$

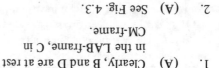

Solutions

1. (A) Clearly, B and D are at rest in the LAB-frame, C in CM-frame.

2. (A) See Fig. 4.3.

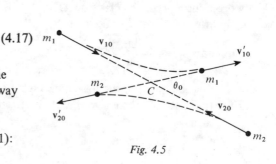

Fig. 4.5

The formulae in the LAB-frame are therefore much more complicated, and that is why nuclear physicists like to make their calculations in the CM-frame. Unfortunately, their measurements are made in the LAB-frame, where the target particle, to which we shall give the suffix 2, is at rest. We must therefore learn how to translate from one frame into the other, and this is done most easily in terms of momenta.

In the CM-frame, the collision proceeds as shown, and each particle is scattered through the same *scattering angle* θ_0, in the LAB-frame the scattering angle θ_1 is measured between the initial and final directions of the incident particle, while the *recoil angle* θ_2 is measured between the initial direction of the incident and the final direction of the target particle.

We obtain the relation between the momenta in the two frames by multiplying the two equations (3.78) by m_1 and m_2 respectively and differentiating with respect to time:

$$\mathbf{p}_1 = m_1 \mathbf{V} + \mathbf{p}_0,$$

$$\mathbf{p}_2 = m_2 \mathbf{V} - \mathbf{p}_0. \tag{4.19}$$

Since $\mathbf{p}_2 = 0$, this leads to

$$\mathbf{p}_1 = \frac{m_1 + m_2}{m_2} \mathbf{p}_0. \tag{4.20}$$

Similarly, after the collision,

$$\mathbf{p}_1' = \frac{m_1}{m_2} \mathbf{p}_0 + \mathbf{p}_0',$$

$$\mathbf{p}_2' = \mathbf{p}_0 - \mathbf{p}_0'. \tag{4.21}$$

This at once leads to the equation for conservation of momentum in the LAB-frame,

$$\mathbf{p}_1 = \mathbf{p}_1' + \mathbf{p}_2'. \tag{4.22}$$

Relations (4.20), (4.21), (4.22) are most easily described by the vector diagram of Fig. 4.7, which also shows the relation between the angles. By dropping a perpendicular from O, and remembering that $p_0 = p_0'$, we find that

$$\tan \theta_1 = \frac{\sin \theta_0}{(m_1/m_2) + \cos \theta_0} \tag{4.23}$$

for the relation between the scattering angles in the two systems. For the relation between the energies we have, from (4.17) and (4.20), that

$$T_0 = \frac{p_0^2}{2\mu},$$

$$T = \frac{p_1^2}{2m_1} = \frac{m_1 + m_2}{m_2} T_0. \tag{4.24}$$

Before   After

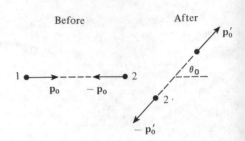

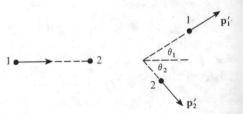

CM-frame

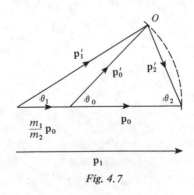

LAB-frame

Fig. 4.6

Fig. 4.7

From the point of view of the effectiveness of a collision process as a means of investigating atomic structure it is the energy in the CM-frame that matters, and this is unfortunately always less than the energy that is produced in the particle accelerator, which is of course in the LAB-frame.

Worked example 4.1. Show that for an incident particle of mass greater than that of the target particle, there is a maximum scattering angle in the LAB-frame. If a deuteron is scattered elastically by a proton, show that this maximum is $30°$ and that the energy transferred to the proton in that case is $\frac{2}{3}$ of the deuteron incident energy. (The mass of the deuteron is twice that of the proton.)

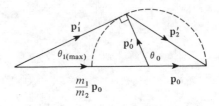

Fig. 4.8

For $m_1 > m_2$, the vector diagram is as shown. Clearly the maximum occurs when $\theta_1 = \theta_0 - 90°$, and since $p_0' = p_0$,

$$\sin \theta_{1(max)} = m_2/m_1.$$

As the mass of the deuteron is twice that of the proton, $\theta_{1\,(max)} = 30°$.
    The transferred energy is

$$T_2' = \frac{p_2'^2}{2m_2}$$

and it follows from Fig. 4.6 quite generally that

$$p_2' = 2p_0 \sin (\theta_0/2).$$

Hence from (4.24), it follows, after some simplification, that

$$\frac{T_2'}{T} = \frac{4m_1 m_2}{(m_1 + m_2)^2} \sin^2 \frac{\theta_0}{2}$$

which reduces to $\frac{2}{3}$ for $\theta_1 = \theta_{1\,(max)}$.

Progress test

1. Kinetic energy in an inelastic collision can be lost to all except which of the following?

    A   Internal energy
    B   Potential energy
    C   Electromagnetic radiation
    D   Sound.

2. The total kinetic energy of two colliding particles, T_0 in the CM-frame and T in the LAB-frame satisfies:

    A $T_0 \geqslant T$

    B $T_0 = T$

    C $T_0 \leqslant T$

    D $T_0 < T$.

3. A particle m_1 can be scattered by a stationary particle m_2 through an angle greater than 90°

    A   always

    B   if $m_1 < m_2$

    C   if $m_1 = m_2$

    D   if $m_1 > m_2$.

4.5 Elastic collisions at relativistic energies

At very high energies, we have to use the relativistic expressions for momentum and energy. Let the velocity of the CM-frame relative to the LAB-frame be V. Then, since the target particle is at rest in the LAB-frame, its velocity in the CM-frame before the collision is

$$\mathbf{v}_{20} = -\mathbf{V} \tag{4.25}$$

and its momentum is

$$\mathbf{p}_{20} = -\frac{m_2 \mathbf{V}}{\sqrt{1 - V^2/c^2}}. \tag{4.26}$$

The CM-frame is again defined (see problem 3.9) as the one in which the system as a whole has zero momentum, and hence

$$\mathbf{p}_{10} = -\mathbf{p}_{20}. \tag{4.27}$$

For the total energies of two particles of rest masses m_1 and m_2, we then have

$$\begin{aligned}
E_{10}^2 &= p_{10}^2 c^2 + m_1^2 c^4, \\
E_{20}^2 &= p_{20}^2 c^2 + m_2^2 c^4.
\end{aligned} \tag{4.28}$$

Hence

$$E_{10}^2 = E_{20}^2 + (m_1^2 - m_2^2) c^4. \tag{4.29}$$

We wish to compare the energies of the incident particle in the CM- and LAB-frames. As all velocities are in the same direction, we can omit the vector notation. From problem 2.10, we have

$$E_1 = \frac{E_{10} + p_{10} V}{\sqrt{1 - V^2/c^2}}. \tag{4.30}$$

We now eliminate V and p_{10} between (4.30), (4.27), (4.26) and the equation

$$E_{20} = \frac{m_2 c^2}{\sqrt{1 - V^2/c^2}} \qquad (4.31)$$

and obtain

$$E_1 m_2 c^2 = E_{10} E_{20} + \sqrt{(E_{10}^2 - m_1^2 c^4)(E_{20}^2 - m_2^2 c^4)}. \quad (4.32)$$

This expression is particularly simple for $m_1 = m_2$, since then $E_{10} = E_{20}$, and (4.32) reduces to

$$E_{10} = \sqrt{\tfrac{1}{2}(E_1 + m_1 c^2) m_1 c^2}. \qquad (4.33)$$

Thus, at very high energies, when $E_1 \gg m_1 c^2$, the energy in the CM-frame increases only as the square root of the energy in the LAB-frame.

4.6 Scattering cross-sections

The rest of this chapter is concerned with the scattering of particles by other particles. This is an extremely important subject, both in itself and because it introduces ideas of probability into classical mechanics. It is also quite difficult, and as nothing in the following chapters depends on it, it may well be omitted at first reading. It should not however be omitted at second reading.

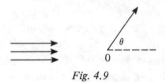

Fig. 4.9

One of the best ways of finding out something about an atom, nucleus or elementary particle is to bombard it with a stream of particles and to measure the number of these deflected by the target particle through a given angle θ, the *scattering angle*. In the treatment of this problem we shall again confine ourselves to velocities for which non-relativistic kinematics are appropriate.

The co-ordinate system best suited to the analysis of scattering experiments is the set of polar co-ordinates, with origin at the target particle and the $\theta = 0$ axis along the direction of the incident beam. The position in space of a point P is then determined by the *radius vector r*, the *co-latitude θ* and the *azimuth* or *longitude φ*. On a sphere of radius r, consider the element $PQRS$ formed by the intersection of the lines for which θ, $\theta + d\theta$, φ and $\varphi + d\varphi$ are respectively constant. Then

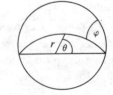

Fig. 4.10

$$PQ = r\,d\theta, \qquad PS = r \sin \theta \, d\varphi, \qquad (4.34)$$

so that the area $PQRS$ is given by

$$PQRS = r^2 \sin \theta \, d\theta \, d\varphi. \qquad (4.35)$$

We next introduce the idea of a *solid angle*. The solid angle subtended by the area S at a point O is given by the area A cut out of a unit sphere, centre O, by the cone with vertex O and base S. This is a natural

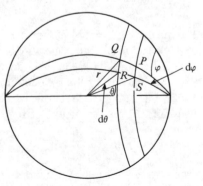

Fig. 4.11

extension of the definition of a plane angle, when measured in radians, and the unit of solid angle is the *steradian* (sr). If we let the area S envelop the point O completely, then A will cover the whole surface of the sphere. Hence it is clear that the solid angle subtended by any closed surface at any point inside it is 4π. Now let the unit vectors along the radius vector $\mathbf{r}$ and perpendicular to the small area dS be denoted by $\hat{\mathbf{r}}$ and $\hat{\mathbf{n}}$ respectively. Then the solid angle subtended at O by dS is

$$d\Omega = \frac{dS}{r^2}\cos\beta, \tag{4.36}$$

where β is the angle between $\hat{\mathbf{r}}$ and $\hat{\mathbf{n}}$. Alternatively, we note that the area has both a magnitude, dS, and defines a direction, $\hat{\mathbf{n}}$. We can therefore specify it by the vector $d\mathbf{S}$ and write

$$d\Omega = \frac{\hat{\mathbf{r}}\cdot d\mathbf{S}}{r^2}. \tag{4.36'}$$

We next turn to the specification of the incident beam. Let it consist of n particles per unit volume, moving with velocity v along the direction $\theta = 0$. Then the *beam flux f* is defined as the number of particles in the beam which cross unit area at right angles to the beam per second. At the beginning of a particular second, all the particles in the beam which will cross the unit area by the end of the second are contained in a cylinder, with the unit area as base and of length v. As this has a volume v and hence contains nv particles,

$$f = nv. \tag{4.37}$$

Let the beam be scattered by an obstacle, the *scattering centre*, which presents a certain cross-sectional area σ to the beam. This area is found by measuring the number w of particles which strike the obstacle in unit time, for unit cross-sectional area of the incident beam. We then have

$$\frac{nv}{1} = \frac{w}{\sigma} \qquad \text{or} \qquad w = f\sigma. \tag{4.38}$$

The quantity σ is the *total scattering cross-section* for the particular obstacle and beam and is measured in m^2.

Now consider the number dw which is scattered by the obstacle S in unit time out of the incident beam of unit cross-sectional area into a solid angle $d\Omega$ in a direction given by the polar angles θ, φ. We can then put, corresponding to (4.38),

$$dw = f\,d\sigma = f\,\frac{d\sigma}{d\Omega}\,d\Omega, \tag{4.39}$$

where we can measure the quantities f, dw and $d\Omega$. In this way we find the *differential scattering cross-section* $d\sigma/d\Omega$. This is, in general, a

Solutions

1. (B) The energies are measured at infinity, where there is no potential energy.

2. (D) See (4.24).

3. (B) For $m_1 > m_2$, θ_1(max) is acute, and for $m_1 = m_2$ it is 90°.

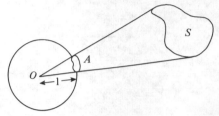

Fig. 4.12

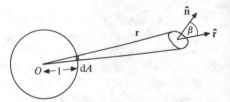

Fig. 4.13

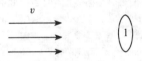

Fig. 4.14

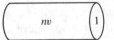

Fig. 4.15

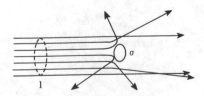

Fig. 4.16

function of the scattering angle θ. (It may also be a function of the azimuthal angle φ, but in many cases the experiment is arranged with cylindrical symmetry about the axis of the incident beam. In this case we say that the experiment has *axial symmetry*, and the differential scattering cross-section is a function of θ only.) From (4.39), we have

$$\frac{d\sigma}{d\Omega} = \frac{dw}{d\Omega} \bigg/ f$$

$$= \frac{\text{number of scattered particles per unit solid angle}}{\text{number of incident particles per unit area}} \qquad (4.40)$$

and the units are $m^2 \, sr^{-1}$. The total cross-section is obtained from the differential cross-section by integrating over all solid angles,

$$\sigma = \int \frac{d\sigma}{d\Omega} \, d\Omega. \qquad (4.41)$$

Note that σ is the size of the obstacle as 'seen' by the incident beam. It is not necessarily the same as the geometric cross-section, as the obstacle may be partially transparent to the beam.

  Finally, we link the scattering cross-section to the trajectories of the particles in the beam. Let a given particle be initially at a distance b, called the *impact parameter*, from the axis through the scattering centre S, and let this particle be scattered through an angle θ into a detector located at polar angles θ, φ and at distance L from S. It is assumed that the detector is effectively at a large distance from S, so that $L \gg b$. Let the detector have area dS, corresponding to small angles $d\theta$ and $d\varphi$ and let the particles that enter dS have impact parameters between b and $b + db$. Then dS subtends a solid angle

$$d\Omega = \frac{dS}{L^2} = \sin\theta \, d\theta \, d\varphi \qquad (4.42)$$

at S, where we have used (4.35) and (4.36). For the particle to enter dS, it must have been in that part of the incident beam that crossed area

$$dB = b|db|d\varphi. \qquad (4.43)$$

If the incident flux is f, then the number of particles scattered into $d\Omega$ in unit time is

$$dw = f \, dB = f \, b|db|d\varphi. \qquad (4.44)$$

Therefore the number of particles scattered into unit solid angle is

$$\frac{dw}{d\Omega} = \frac{f \, b|db|d\varphi}{\sin\theta \, d\theta \, d\varphi}, \qquad (4.45)$$

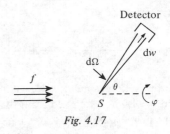

Fig. 4.17

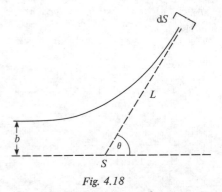

Fig. 4.18

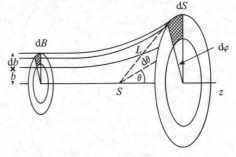

Fig. 4.19

so that, using (4.40)

$$\frac{d\sigma}{d\Omega} = \frac{b|db|}{\sin\theta\, d\theta} = \frac{b}{\sin\theta}\left|\frac{db}{d\theta}\right|. \tag{4.46}$$

In order to determine $d\sigma/d\Omega$, we therefore require the relation between b and θ, which is obtained from the trajectory. This, in turn, depends on the properties of the scattering centre and scattered particle, and in the next two sections we shall solve two particular scattering problems. The reason for writing down the absolute value of $db/d\theta$ in (4.46) is that this quantity can be negative, but a cross-section is always positive. It may also be noticed that, once b has been expressed in terms of θ in (4.46), the cross-section is given purely in terms of the scattering angle, and does not depend on the paths of the individual scattered particles.

We have, of course, implicitly made the assumption that the particles in the incident beam are distributed uniformly across the beam, and that they do not interact with each other. To that extent our treatment is based on statistical averaging and therefore depends on there being a large number of particles in the beam. For a single particle in the beam, the differential scattering cross-section for a particular angle is proportional to the probability of the particle being scattered through θ.

Progress test

1. The largest possible value of (a) the co-latitude θ, (b) the longitude φ is

    A $\frac{1}{2}\pi$
    B π
    C $\frac{3}{2}\pi$
    D 2π.

2. The solid angle substended by the surface of the cube at (a) X, the midpoint of the cube, (b) Y, the midpoint of the face $PQRS$, (c) Z, the midpoint of the edge PQ, and (d) P, a corner of the cube is

    A $\frac{1}{2}\pi$
    B π
    C 2π
    D 4π.

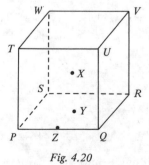

Fig. 4.20

3. The unit in which the differential scattering cross-section is measured is

    A m sr
    B m^2 sr
    C m sr^{-1}
    D   none of the above.

4.7   Scattering by a hard sphere

We first consider scattering by a perfectly elastic sphere of radius R.
Such a sphere has the property that all particles hitting it suffer
perfectly elastic collisions.

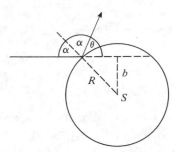

It follows from the figure that

$$b = R \sin \alpha = R \cos (\theta/2). \tag{4.47}$$

We substitute in (4.46) and obtain

$$\frac{\mathrm{d}\sigma}{\mathrm{d}\Omega} = \frac{R \cos (\theta/2)}{\sin \theta} \cdot \tfrac{1}{2} R \sin \frac{\theta}{2} = \tfrac{1}{4} R^2. \tag{4.48}$$

In this special case the differential scattering cross-section is independent
of θ, i.e., the scattering is the same in all directions. It is called *isotropic*.
The total scattering cross-section is obtained from (4.41),

$$\sigma = \int \tfrac{1}{4} R^2 \, \mathrm{d}\Omega = \tfrac{1}{4} R^2 \int \mathrm{d}\Omega = \tfrac{1}{4} R^2 \, 4\pi.$$

$$\therefore \quad \sigma = \pi R^2. \tag{4.49}$$

In this case the total scattering cross-section is equal to the geometric
cross-section, as one would expect.

Fig. 4.21

4.8   Rutherford scattering

Our other example concerns the famous experiment, due to Rutherford
(1871–1937), of the scattering of α-particles, of charge q and mass m, from
an atomic nucleus of charge q', the mass of which is taken to be infinite.
(If the mass is taken to be finite, then we must replace m by the reduced
mass (4.14).) The force between the α-particle and the nucleus, when
the two are a distance r apart, is the Coulomb force

$$\mathbf{F} = \frac{K}{r^2} \hat{\mathbf{r}}, \quad \text{where} \quad K = \frac{qq'}{4\pi\epsilon_0}. \tag{4.50}$$

Let the particle start at infinity at A with momentum $\mathbf{p}_A$ in direction
$\overrightarrow{AS}$ and end at infinity at B with momentum $\mathbf{p}_B$ in direction $\overrightarrow{SB}$. As the
scattering is elastic, $p_A = p_B = p$, say. Let the change in momentum be

$$\Delta\mathbf{p} = \mathbf{p}_B - \mathbf{p}_A,$$

that is

$$\Delta\mathbf{p} = 2p \sin \frac{\theta}{2} \quad \text{along } \overrightarrow{SN}, \tag{4.51}$$

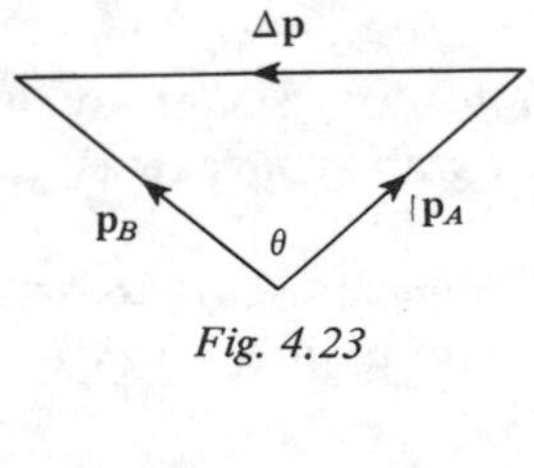

where $\overrightarrow{SN}$ bisects $\angle ASB$. We require a relation between the impact
parameter b and the scattering angle θ.

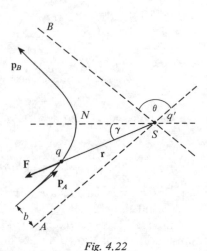

Fig. 4.22

Fig. 4.23

94

[4.8

Now the change of momentum must be due to the component of **F** along $\overrightarrow{SN}$ acting during the time of interaction:

$$\Delta p = \int_A^B F \cos\gamma \, dt = \int_A^B \frac{K}{r^2} \cos\gamma \, dt. \qquad (4.52)$$

Also, the angular momentum about S is conserved. If v is the incident velocity at infinity, then equating the angular momentum when the α-particle is at A and at P (see (3.16)), we have

$$mvb = mr^2 \dot{\gamma}. \qquad (4.53)$$

Substituting for r^2 in (4.52), we obtain

$$\Delta p = \int_A^B \frac{K}{vb} \dot{\gamma} \cos\gamma \, dt$$

$$= \int_{-(\pi-\theta)/2}^{(\pi-\theta)/2} \frac{K}{vb} \cos\gamma \, d\gamma = \frac{2K}{vb} \cos\frac{\theta}{2},$$

so that, from (4.51),

$$2mv \sin\frac{\theta}{2} = \frac{2K}{vb} \cos\frac{\theta}{2},$$

or

$$b = \frac{K}{mv^2} \cot\frac{\theta}{2}. \qquad (4.54)$$

This is the required relation, which is now substituted in (4.46), to give

$$\frac{d\sigma}{d\Omega} = \left(\frac{qq'}{4\pi\epsilon_0 \, mv^2}\right)^2 \frac{1}{4\sin^4(\theta/2)}. \qquad (4.55)$$

This is Rutherford's famous formula. [For its application see Jackson, chapter 4 and also chapter 7.]

Although we derived this formula classically, as did Rutherford, it turns out that it is also valid when derived from quantum theory, which is fortunate since the nucleus and α-particle are in fact a quantum system. It is the only scattering cross-section that yields the same result in the two theories, and the only one that can be evaluated classically without an explicit calculation of the trajectories.

Worked example 4.2. An α-particle of energy 2×10^{-13} J is scattered by an aluminium atom through an angle of $90°$. Calculate the distance of

Solutions

1. (a) (B), (b) (D).

2. (a) (D), (b) (C), (c) (B), (d) (A).
    This is most easily seen by considering the set of eight adjacent cubes that have a common corner P.

3. (D)   The unit is m^2 sr^{-1}.

*closest approach. [The charges on the α-particle and the aluminium atom
are 2e and 13e, where e = 1·60 × 10⁻¹⁹ coulomb is the proton charge;
$4\pi\epsilon_0 = 10^7/c^2$, c = 3 × 10⁸ m s⁻¹ is the velocity of light.]*

We apply the conservation laws of energy and angular momentum
to the situation at N and A:

$$\tfrac{1}{2}mv^2 + \frac{2Ze^2}{4\pi\epsilon_0 d} = \tfrac{1}{2}mv_0^2,$$

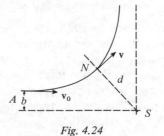

Fig. 4.24

where $Z = 13$ for Al, and

$$vd = v_0 b.$$

Also, from (4.54),

$$b = \frac{2Ze^2}{4\pi\epsilon_0 mv^2}.$$

We put $2Ze^2/(4\pi\epsilon_0) = K$ and eliminate v and b:

$$v = \frac{K}{mv_0 d}.$$

$$\therefore \quad \tfrac{1}{2}m\frac{K^2}{m^2 v_0^2 d^2} + \frac{K}{d} = \tfrac{1}{2}mv_0^2.$$

$$\therefore \quad d^2 - \frac{2K}{mv_0^2}d = \left(\frac{K}{mv_0^2}\right)^2.$$

$$\therefore \quad d = (1 + \sqrt{2})\frac{K}{mv_0^2}.$$

$$K = \frac{2 \times 13 \times (1\cdot6 \times 10^{-19})^2}{10^7/9 \times 10^{16}}$$

$$= 3 \times 10^{-27}\ \text{J m},$$

$$\tfrac{1}{2}mv_0^2 = 2 \times 10^{-13}\ \text{J}.$$

$$\therefore \quad d = 3\cdot6 \times 10^{-14}\ \text{m}.$$

Progress test

1. In Rutherford scattering, the probability of a particle being
    scattered through a particular angle θ.

    A   is the same for all θ
    B   increases with θ
    C   decreases as θ increases
    D   increases up to $\theta = 90°$ and then decreases.

4.9 Mean free path in matter

As a different illustration of collision theory, we consider the motion
of a particle through matter, say the motion of an electron through a
medium containing n atoms per unit volume. Let the total cross-section
for scattering by one atom be σ. We can think of this as a point particle
meeting a lot of atoms, each of cross-sectional area σ, or alternatively
as a particle of cross-sectional area σ meeting a lot of point atoms. In the
latter view, the particle in moving through a distance l sweeps out a
cylinder of volume σl and in the process collides with $n\sigma l$ atoms. Its
average distance between collisions, the *mean free path*, is therefore

$$\lambda = \frac{l}{n\sigma l} = \frac{1}{n\sigma}. \tag{4.56}$$

Note that, here again, we are using statistical considerations, and that
we cannot determine the individual free paths between collisions from
the information available.

　　We can use the idea of the mean free path to derive an expression
for the attenuation of a beam of particles that is absorbed in passing
through matter. Let the beam, of flux $f(0)$, impinge normally on unit
area of the front face of the absorber, and denote the flux of particles
which penetrate distance x into the absorber without collisions by $f(x)$.
In a further slice of absorber, thickness dx, there are $n\,dx$ atoms and
these present a total cross-sectional area $n\sigma\,dx$ to each incident
particle. Hence the number of particles that strike atoms in the slice in
unit time (see (4.38)) is $n\sigma f(x)\,dx$ and this is the reduction in the flux
through absorption in the slice, that is,

$$f(x) - f(x + dx) = n\sigma f(x)\,dx. \tag{4.57}$$

$$\therefore \quad \frac{df}{dx} = -n\sigma f.$$

This can be integrated, to give

$$f(x) = f(0)e^{-n\sigma x} = f(0)e^{-x/\lambda}, \tag{4.58}$$

using (4.56). Thus the attenuation is exponential, and the flux is reduced
to e^{-1} times its incident value in one mean free path.

　　To calculate the mean free path of a molecule in a gas, we have to
take account of the fact that, now, neither the incident particle nor the
obstacles can be treated as points, since they are all like molecules. Now,
two molecules collide if their centres are less than a molecular diameter
d apart. Hence, the cylinder swept out by one molecule, within which

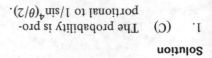

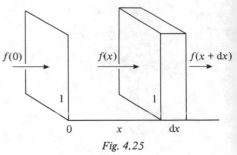

$f(0) \quad\quad f(x) \quad\quad f(x + dx)$

Fig. 4.25

Fig. 4.26

are the centres of molecules that cause collisions, has radius d. The collision cross-section is therefore

$$\sigma = \pi d^2. \tag{4.59}$$

Progress test

1. On which one of the following does the absorption of a beam of particles passing through an absorbing material not depend?

    A   The thickness of the absorber
    B   The mass of the particles of the absorber
    C   The number of particles of the absorber per unit volume
    D   The scattering cross-section for each incident particle by an absorber particle.

4.10 Problems

4.1. A bomb of mass 100 kg is dropped from an aeroplane. After falling 500 m it explodes in mid-air into two fragments of 40 kg and 60 kg mass. If the disintegration energy is 120 000 J, show that the maximum velocity relative to the aeroplane of the larger fragment immediately after the explosion is 140 m s^{-1}.

4.2. Show that if a particle is scattered by another particle of the same mass, then the angle between the directions of motion of the scattered and recoiling particles in the LAB frame is always 90°, and that the maximum scattering angle in the LAB frame is also 90°.

4.3. A neutron of mass m_1 collides elastically with a nucleus in moderator material of mass m_2. Show that the kinetic energy T_1' of the neutron in the LAB frame after the collision satisfies the inequality.

$$\left(\frac{m_2 - m_1}{m_2 + m_1}\right)^2 T_1 \leqslant T_1' \leqslant T_1,$$

where T_1 is the kinetic energy of the neutron before the collision.

4.4. A proton scatters elastically from a particle in a bubble chamber, the scattering and recoil angles being θ_1 and θ_2. How would you use Fig. 4.7 to find the ratio of the mass of the unknown particle to that of the proton? If $\theta_1 = 63°$ and $\theta_2 = 45°$, show that the unknown particle has twice the mass of a proton.

4.5. The energy of the CERN proton synchrotron is 30 mc^2, where m is the rest mass of the proton. What is the corresponding energy

in the CM frame for proton–proton scattering? What would be the
energy in the CM frame for a LAB frame energy of 300 mc^2? [See
R. R. Wilson, 'Particle Accelerators', *SA* 251, March 1958,
p. 64.]

*4.6. Use (4.33) to show that the incident velocity v_1 and the centre of
mass velocity V, both in the LAB frame, are related by

$$\frac{2}{1 - V^2/c^2} = 1 + \frac{1}{\sqrt{1 - v_1^2/c^2}}$$

What does this expression reduce to in

(a) the extreme relativistic limit, $v_1 \rightarrow c$,
(b) the non-relativistic limit, $v_1 \ll c$.

4.7. Show that the polar co-ordinates of a point with Cartesian co-
ordinates (x, y, z) are given by

$$x = r \sin \theta \cos \varphi, \qquad y = r \sin \theta \sin \varphi, \qquad z = r \cos \theta,$$

and hence express r, θ, φ explicitly in terms of x, y, z. Use the
resulting expressions to find the polar co-ordinates of the points
whose Cartesian co-ordinates are

(a) $(1, 1, 1)$,   (b) $(1, -1, 0)$,
(c) $(1, 0, -1)$, (d) $(0, 0, 1)$.

*4.8. By integrating (4.55) from a certain angle θ_0 to $\frac{1}{2}\pi$, show that the
number of α-particles scattered through angles greater than θ_0 is
proportional to $\cot^2(\theta/2)$. Hence show that the total cross-
section for α-particles being scattered by an atomic nucleus is
infinite.

*4.9. A paraboloid of rotation is formed by rotating the parabola

$$y^2 = 4ax$$

about the x-axis. A beam of particles parallel to the x-axis is
scattered elastically by the surface of the paraboloid. Find the
differential cross-section and comment on the result obtained.
[Hint: It helps to express the parabola in parametric co-ordinates,
$x = au^2, y = 2au.$]

4.10. A 1 MeV γ-ray has its intensity reduced by a factor e^{-1} in passing
through 6·10 cm of Al. What is the scattering cross-section for
one atom? Comment on the result. [The number of Al atoms in
a kg of Al is 2·23 x 10^{25}, and the density of Al, $\rho = 2700$ kg m^{-3}.]

5. Systems of particles

5.1 Introduction

So far, we have considered the motion of mechanical systems consisting at most of two particles, and we have found exact solutions to the problem of their motions. Once a system consists of more than two particles, this is no longer possible, although high-speed computers nowadays can give very accurate numerical solutions to the problem of the motion of three bodies, such as the motion of earth, moon and rocket in the Apollo moon landing programme. Beyond that even computers cannot go.

For this reason, we now consider how far we can go algebraically towards a solution of the problem of the motion of any number of particles under mutual internal and also external forces. We shall be able to show that this very complicated motion can be divided into two parts:

(a) the motion of the particles relative to the centre of mass of the whole system, which for this purpose may be considered to be at rest;

(b) the motion of the centre of mass, acting like a particle with a mass equal to the sum of the masses of all the particles in the system.

In order to do this, we shall employ a powerful mathematical notation, which enables us by means of suffixes to handle an indefinite number of particles concisely. If this notation is unfamiliar, then it is a good idea to write out the formulae in full for two or three particles, as we show below.

We consider N particles. In our notation each particle will be labelled by a subscript letter, so that the ith particle has mass m_i and position vector r_i. The force on the ith particle due to external causes will be denoted by F_i and that due to its interaction with the jth particle by F_{ji}. Thus, if the system consists of three particles, the second particle has mass m_2 and position vector r_2, and is acted on by an external force F_2 and internal forces F_{12} and F_{32}, etc. Often, we shall have to sum a property, such as the total mass of the system which

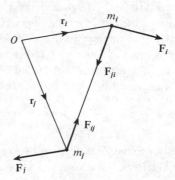

Fig. 5.1

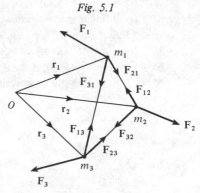

Fig. 5.2

is the sum of the masses of the individual particles, and this will be denoted by the symbol

$$\sum_{i=1}^{N} \cdot$$

For instance, the total mass M of a system of N particles is

$$M = \sum_{i=1}^{N} m_i, \qquad (5.1)$$

which for three particles can be written in full

$$M = \sum_{i=1}^{3} m_i = m_1 + m_2 + m_3.$$

The origin of our inertial reference frame is denoted by O, but we shall also use a second reference frame, not necessarily inertial, in motion relative to the first, and the origin of this will be denoted by O'. In fact, all quantities referred to this second frame will be distinguished through having primes $'$. The centre of mass of the system will be labelled C, and the position vectors of O' and C relative to O will be denoted by s and $\mathbf{R}$ respectively. Note that for the position vector of the ith particle,

$$\mathbf{r}_i = \mathbf{r}_i' + \mathbf{s}. \qquad (5.2)$$

Finally we shall denote quantities referring to the system as a whole by capital letters. We already used that notation in (5.1). Thus we shall put

    M = total mass of system,
    $\mathbf{P}$ = total linear momentum of system,
    $\mathbf{L}$ = total angular momentum of system about O,
    T = total kinetic energy of system,
    $\mathbf{F}$ = resultant force on system,
    $\mathbf{N}$ = resultant torque on system about O.

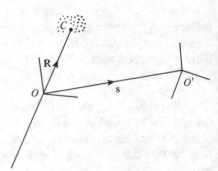

Fig. 5.3

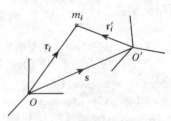

Fig. 5.4

Progress test

1. $\mathbf{F}_{j1} + \mathbf{F}_{j2} + \mathbf{F}_{j3} + \cdots + \mathbf{F}_{jN}$ can be written

    A $\sum_{j=1}^{N} \mathbf{F}_{ij}$

B $\quad \sum\limits_{i=1}^{N} \mathbf{F}_{ij}$

C $\quad \sum\limits_{k=1}^{N} \mathbf{F}_{jk}$

D $\quad \sum\limits_{i=1}^{N} \sum\limits_{j=1}^{N} \mathbf{F}_{ij}.$

2. If $\sum\limits_{i=1}^{3} x_i = 0$, then

A $\quad x_1 + x_2 = x_3$
B $\quad x_1 + x_2 = -x_3$
C $\quad x_1 - x_2 = x_3$
D $\quad$ none of these.

5.2 Motion of the centre of mass

We begin by applying Newton's second law to the motion of the ith particle,

$$\mathbf{F}_i + \sum\limits_{j=1}^{N} \mathbf{F}_{ji} = m_i \ddot{\mathbf{r}}_i. \tag{5.3}$$

Here we can have $i = 1, 2, 3, \ldots, N$. [If we had a system of three particles, the equation of motion of the second particle would be

$$\mathbf{F}_2 + \mathbf{F}_{12} + \mathbf{F}_{22} + \mathbf{F}_{32} = m_2 \ddot{\mathbf{r}}_2.$$

As the particle does not exert a force on itself, we put $F_{22} = 0$, and in general $F_{ii} = 0$. [This is the last time that we shall illustrate our equations by writing them out in full for three particles, but the reader is advised to continue doing so until he is thoroughly familiar with the notation.]

Next, we consider all N equations (5.3) for $i = 1, 2, \ldots, N$ and add these. This gives

$$\sum\limits_{i=1}^{N} \mathbf{F}_i + \sum\limits_{i=1}^{N} \sum\limits_{j=1}^{N} \mathbf{F}_{ji} = \sum\limits_{i=1}^{N} m_i \ddot{\mathbf{r}}_i. \tag{5.4}$$

Now by Newton's third law, the internal forces are equal and opposite in pairs, i.e.

$$\mathbf{F}_{ij} = -\mathbf{F}_{ji} \qquad \text{for all } i, j, \tag{5.5}$$

so that all the terms in the double summation in (5.4) cancel in pairs. The first sum just yields the resultant of all the external forces,

$$\mathbf{F} = \sum_{i=1}^{N} \mathbf{F}_i. \tag{5.6}$$

Finally, in a simple and obvious extension of (3.71) we define the centre of mass position vector R through

$$M\mathbf{R} = \sum_{i=1}^{N} m_i \mathbf{r}_i, \tag{5.7}$$

where

$$M = \sum_{i=1}^{N} m_i.$$

We differentiate twice and substitute in (5.4) to obtain

$$\mathbf{F} = M\ddot{\mathbf{R}}. \tag{5.8}$$

This is the equation of motion of a particle of mass M, situated at $\mathbf{R}$ under a force $\mathbf{F}$, i.e., the motion of the centre of mass C of the system is given by that of a particle with the total mass of the system concentrated at the centre of mass C under the resultant force on the system acting at C. If there is no resultant external force, then

$$M\ddot{\mathbf{R}} = 0,$$

and the centre of mass is at rest or in uniform motion.

In the following sections we shall investigate how the kinetic energy, linear momentum and angular momentum of the system are

Solutions

1. (C) Note that we could use another letter instead of k, but not of j. For that reason k is called a *dummy* suffix.

2. (B) This should be obvious.

divided between the motion of the total mass concentrated at the centre of mass and the motion of the particles of the system relative to their centre of mass.

Progress test

Consider a system of four particles of masses 2, 3, 5 and 10 kg momentarily at the points shown with velocities indicated by single arrows and under forces indicated by double arrows. (Co-ordinate distances are in m.)

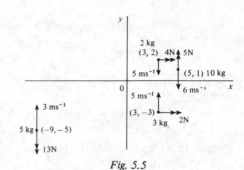

Fig. 5.5

Now answer the following questions:

1. If $\sin 37° = 0.6$, the angle which the resultant of the four forces makes with Ox is

    A $37°$
    B $53°$
    C $127°$
    D $307°$.

2. The magnitude, in N, of the resultant force is

    A   6
    B   10
    C   20
    D   none of these.

3. The centre of mass of the system has co-ordinates

    A $(1, 1)$
    B $(-1, 1)$
    C $(1, -1)$
    D $(-1, -1)$.

4. The velocity of the centre of mass has components, in m s^{-1},

    A $(6, -2)$
    B $(0.6, -2)$
    C $(0, -2)$
    D $(0, -3)$.

5.3   Kinetic energy

The kinetic energy of the system relative to O is obtained by summing over the kinetic energies of all the particles,

$$T_O = \tfrac{1}{2} \sum m_i \dot{\mathbf{r}}_i^2.$$ (5.9)

Relative to O', we have

$$T_{O'} = \tfrac{1}{2} \sum m_i (\dot{\mathbf{r}}_i - \dot{\mathbf{s}})^2$$

$$= \tfrac{1}{2} \sum m_i (\dot{\mathbf{r}}_i^2 - 2\dot{\mathbf{r}}_i \cdot \dot{\mathbf{s}} + \dot{\mathbf{s}}^2)$$

$$= \tfrac{1}{2} \sum m_i \dot{\mathbf{r}}_i^2$$

$$- (\sum m_i \dot{\mathbf{r}}_i) \cdot \dot{\mathbf{s}} + \tfrac{1}{2}(\sum m_i) \dot{\mathbf{s}}^2$$

$$= T_O - M\dot{\mathbf{R}} \cdot \dot{\mathbf{s}} + \tfrac{1}{2} M\dot{s}^2,$$ (5.10)

using (5.9), (5.7) and (5.1). If O' is C, then we have

$$T_C = T_O - \tfrac{1}{2} M\dot{R}^2$$ (5.11)

or, on transposing,

K.E. relative to O = K.E. relative to C

+ K.E. of total mass at C. (5.12)

In this way we have split up the kinetic energy into its two parts, as was our intention.

Progress test

The following questions refer to the system of particles in Fig. 5.5.

1.   The kinetic energy of the system, in joules, relative to 0 is

   A   95
   B   150
   C   315
   D   none of these.

2.   The kinetic energy of the system, in joules, relative to the centre of mass is

   A   0
   B   225
   C   305
   D   450.

Solutions

1.   (D)   The resultant has components (6, −8).

2.   (B) $\sqrt{(6^2 + 8^2)} = 10$.

3.   (C) $20\bar{x} = 2 \times 3 + 10 \times 5 + 3 \times 3 + 5 \times (−9) = 20$, etc.

4.   (C) $20\bar{v}_y = 2 \times (−5) + 10 \times (−6) + 3 \times 5 + 5 \times 3 = −40$.

5.4 Linear momentum

The linear momentum of the system relative to O is

$$\mathbf{P}_O = \sum m_i \dot{\mathbf{r}}_i = M\dot{\mathbf{R}} \tag{5.13}$$

from (5.7). Hence we can write (5.8) as

$$\dot{\mathbf{P}}_O = \mathbf{F}, \tag{5.14}$$

so that Newton's second law applies to the motion of the centre of mass.

To obtain the linear momentum relative to O', the origin of our second frame of reference, we use (5.2) and write

$$\mathbf{P}_{O'} = \sum m_i \dot{\mathbf{r}}_i'$$

$$= \sum m_i (\dot{\mathbf{r}}_i - \dot{\mathbf{s}}).$$

[Note that we often do not write the limits $i = 1$ and $i = N$ on the sum, when these are obvious.] Now

$$\sum m_i \dot{\mathbf{s}} = \left(\sum m_i \right) \dot{\mathbf{s}},$$

since it is permissible to take the quantity $\dot{\mathbf{s}}$, which is the same for all terms of the sum, outside the summation symbol. Hence, using (5.13) and (5.1),

$$\mathbf{P}_{O'} = \mathbf{P}_O - M\dot{\mathbf{s}}. \tag{5.15}$$

If O' is C, then s must be replaced by $\mathbf{R}$ and we have

$$\mathbf{P}_C = \mathbf{P}_O - M\dot{\mathbf{R}} = 0, \tag{5.16}$$

by (5.13). Thus, the linear momentum of a system relative to its centre of mass is zero, a result which, for two particles, we already derived in (4.15). For linear momentum, the splitting up of the motion therefore leads to

Momentum relative to O = zero +

momentum of total mass at C relative to O. (5.17)

Differentiating (5.15) with respect to time, we have

$$\dot{\mathbf{P}}_{O'} = \dot{\mathbf{P}}_O - M\ddot{\mathbf{s}},$$

that is,

$$\mathbf{F}' = \mathbf{F} - M\ddot{\mathbf{s}}, \tag{5.18}$$

where we have applied Newton's second law to motion relative to O'. If O' is moving with uniform velocity relative to O, i.e., if O' also determines an inertial reference frame, then $\ddot{s} = 0$ and the same forces are measured in the two frames of reference. If, on the other hand, O' moves non-uniformly, then we see from (5.18) that its motion introduces the inertial force $M\ddot{s}$.

Solutions

1. (D) $\frac{1}{4}(2 \times 25 + 10 \times 36 + 3 \times 25 + 5 \times 9) = 265.$

2. (B) $265 - \frac{2}{3} \times 20 \times 4 = 225.$

Progress test

The following questions refer to the system of particles in Fig. 5.5.

1. The momentum of the system has components, in MKS units,

    A   (0, 0)
    B   (0, 40)
    C   (60, − 40)
    D   none of these.

2. The momentum of the system relative to its centre of mass is

    A   (0, 0)
    B   (0, 40)
    C   (60, − 40)
    D   none of these.

5.5 Angular momentum

Using the expression (3.15) for the angular momentum of a single particle, we have for the angular momentum of the system about O,

$$\mathbf{L}_O = \sum \mathbf{r}_i \times m_i \dot{\mathbf{r}}_i \tag{5.19}$$

and about O'

$$\begin{aligned}
\mathbf{L}_{O'} &= \sum (\mathbf{r}_i - \mathbf{s}) \times m_i(\dot{\mathbf{r}}_i - \dot{\mathbf{s}}) \\
&= \sum \mathbf{r}_i \times m_i \dot{\mathbf{r}}_i - \mathbf{s} \times \sum m_i \dot{\mathbf{r}}_i \\
&\quad - (\sum m_i \mathbf{r}_i) \times \dot{\mathbf{s}} + (\sum m_i)\mathbf{s} \times \dot{\mathbf{s}} \\
&= \mathbf{L}_O - M\mathbf{s} \times \dot{\mathbf{R}} - M\mathbf{R} \times \dot{\mathbf{s}} + M\mathbf{s} \times \dot{\mathbf{s}}. \tag{5.20}
\end{aligned}$$

This expression simplifies considerably when O' is replaced by C,

$$\mathbf{L}_C = \mathbf{L}_O - M\mathbf{R} \times \dot{\mathbf{R}} = \mathbf{L}_O - \mathbf{R} \times \mathbf{P}_O \tag{5.21}$$

or, on transposing,

A.M. about O = A.M. about C + A.M. of

        total mass at C about O.   (5.22)

Differentiating (5.19), we obtain

$$\dot{\mathbf{L}}_O = \sum \mathbf{r}_i \times m_i \ddot{\mathbf{r}} + \sum \dot{\mathbf{r}}_i \times m_i \dot{\mathbf{r}}_i$$

$$= \sum_i \mathbf{r}_i \times \left(\mathbf{F}_i + \sum_j \mathbf{F}_{ji} \right) + 0, \tag{5.23}$$

using (5.3) and noting that $\dot{\mathbf{r}}_i \times \dot{\mathbf{r}}_i = 0$. We have labelled the summation symbols in this case to avoid confusion.

Now the double sum

$$\sum_i \left(\mathbf{r}_i \times \sum_j \mathbf{F}_{ji} \right) = \sum_i \sum_j \mathbf{r}_i \times \mathbf{F}_{ji}$$

contains terms which can be paired off as

$$(\mathbf{r}_i - \mathbf{r}_j) \times \mathbf{F}_{ji}$$

since $\mathbf{F}_{ji} = - \mathbf{F}_{ij}$. Clearly $\mathbf{r}_i - \mathbf{r}_j$ is a vector in the same direction as $\mathbf{F}_{ji}$, so that

$$(\mathbf{r}_i - \mathbf{r}_j) \times \mathbf{F}_{ji} = 0. \tag{5.24}$$

Hence, (5.23) reduces to

$$\dot{\mathbf{L}}_O = \mathbf{N}_O \tag{5.25}$$

where

$$\mathbf{N}_O = \sum \mathbf{r}_i \times \mathbf{F}_i, \tag{5.26}$$

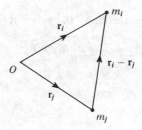

Fig. 5.6

which is the moment of the forces about O, and is called the *torque* of the external forces about O. What (5.24) expresses is that the torque of the internal forces about O is zero. Equation (5.25) is identical in structure to (5.14), and is often referred to as the 'rotational equivalent of Newton's second law' or the 'rotational equation of motion'.

It may be noted that the external forces can have a torque about O, even though there is no resultant force. This happens when the resultant of the external forces yields two equal and opposite forces, not along the same line of action, known as a *couple*. The torque about O of such a couple $\mathbf{F}$, acting at P and $-\mathbf{F}$ at Q, is

$$\mathbf{N} = \mathbf{r}_P \times \mathbf{F} + \mathbf{r}_Q \times (-\mathbf{F})$$

$$= (\mathbf{r}_P - \mathbf{r}_Q) \times \mathbf{F}$$

$$= \mathbf{b} \times \mathbf{F} \tag{5.27}$$

where $\mathbf{b}$ is a vector joining any point Q on the line of action of $-\mathbf{F}$ to any point P on the line of action of $\mathbf{F}$. It is independent of which

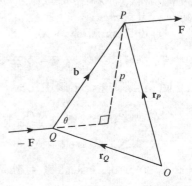

Fig. 5.7

particular points P and Q are chosen, and is also independent of O. The torque of a couple is therefore the same about all points in space. Its magnitude is

$$N = bF\sin\theta = pF \tag{5.28}$$

where θ is the angle between $\mathbf{b}$ and $\mathbf{F}$, and p is the perpendicular distance between the lines of action of $-\mathbf{F}$ and $\mathbf{F}$.

Returning to (5.21), we obtain, by differentiating with respect to time,

$$\dot{\mathbf{L}}_C = \dot{\mathbf{L}}_O - M\dot{\mathbf{R}} \times \dot{\mathbf{R}} - M\mathbf{R} \times \ddot{\mathbf{R}}$$

$$= \dot{\mathbf{L}}_O - \mathbf{R} \times \mathbf{F}. \tag{5.29}$$

Now the torque of the external forces about C is, from (5.26),

$$\mathbf{N}_C = \sum (\mathbf{r}_i - \mathbf{R}) \times \mathbf{F}_i$$

$$= \mathbf{N}_O - \mathbf{R} \times \mathbf{F}. \tag{5.30}$$

Hence

$$\dot{\mathbf{L}}_C = \mathbf{N}_C. \tag{5.31}$$

Thus, the rotational equation of motion takes the same form, whether referred to C or to O, and this is true even when C does not move uniformly relative to O, i.e., when C defines a non-inertial frame of reference. This result, that we may use the simple equation (5.31) in referring the motion of a system of particles to its centre of mass, independently of whether the latter is moving uniformly in an inertial reference frame or not, greatly simplifies the solution of problems in rotational motion.

Worked example 5.1. Show that the equation $\dot{\mathbf{L}} = \mathbf{N}$ is valid in all inertial frames of reference. Discuss modifications to the expression in non-inertial frames.

On differentiating (5.20) with respect to time, we have

$$\dot{\mathbf{L}}_{O'} = \dot{\mathbf{L}}_O - M\mathbf{s} \times \ddot{\mathbf{R}} - M\mathbf{R} \times \ddot{\mathbf{s}} + M\mathbf{s} \times \ddot{\mathbf{s}}$$

since all other terms cancel. Hence

$$\dot{\mathbf{L}}_{O'} = \mathbf{N}_O - \mathbf{s} \times \mathbf{F} - M(\mathbf{R} - \mathbf{s}) \times \ddot{\mathbf{s}}$$

$$= \mathbf{N}_{O'} - M(\mathbf{R} - \mathbf{s}) \times \ddot{\mathbf{s}}$$

using (5.30) with s instead of $\mathbf{R}$. The last term vanishes if

(i) $\ddot{\mathbf{s}} = 0$, i.e., O' is the origin of an inertial frame,

Solutions

1. (D) The answer is $(0, -40)$.

2. (A) The momentum of any system relative to its centre of mass is zero.

(ii)   s = **R**, i.e., O' is the centre of mass of the system, or

(iii) $\ddot{s} \parallel \mathbf{R} - s$, i.e., The acceleration of O' is along $O'C$.

In all other cases there is an *inertial torque* of the inertial force

$$\mathbf{F'} = M\ddot{\mathbf{s}}$$

[see (5.18)] acting at C.

Progress test

The following questions refer to the system of particles in Fig. 5.5.

1.   The angular momentum about the origin of the system, in MKS units, is

    A   − 57

    B   + 420

    C   − 420

    D   none of these.

2.   The torque of the forces about O in MKS units is

    A   − 140

    B   140

    C   156

    D   none of these.

3.   The angular momentum of the system about the centre of mass is

    A   − 460

    B   − 380

    C   0

    D   380.

5.6   Motion relative to centre of mass

Because of the importance of the formulae derived in this chapter we summarize our results below, which can be expressed in the statement:

    Motion of system relative to O =

                Motion of system relative to C

                   + Motion of total mass of system at C.

The explicit formulae are:

    Motion of centre of mass

$$\mathbf{F} = M\ddot{\mathbf{R}} \tag{5.8}$$

Kinetic energy

$$T_O = T_C + \tfrac{1}{2}M\dot{R}^2 \tag{5.11}$$

Linear momentum

$$P_O = M\dot{R} \tag{5.13}$$

$$P_C = O \tag{5.16}$$

$$\dot{P}_O = F. \tag{5.14}$$

Angular momentum

$$L_O = L_C + R \times P_O \tag{5.21}$$

$$\dot{L}_O = N_O \tag{5.25}$$

$$\dot{L}_O = \dot{L}_C + R \times F \tag{5.29}$$

$$\dot{L}_C = N_C \tag{5.31}$$

Solutions

1. (C) $2 \times 3 \times (-5) + 10 \times 5 \times (-6)$
$+ 3 \times 3 \times 5 + 5 \times (-9) \times 3$
$= -420.$

2. (B) $-2 \times 4 + 5 \times 5 - (-) \times (-3) \times 2$
$+ (-) \times (-13) = 140.$

3. (B) $(0, 0, -420) - (1, -1, 0)$
$\times (0, -40, 0) = (0, 0, -380).$
[See ex. 3.11].

Progress test

1. If the resultant of the forces acting on a system of particles is zero, which one of these statements is always true?

   A   The centre of mass of the system is at rest
   B   The particles all move with uniform velocity
   C   The momentum of the system is a minimum
   D   None of the above.

2. Which one of the statements below is correct? The kinetic energy of a system of particles is

   A   the same in all inertial frames of reference
   B   greatest in the frame in which the centre of mass is at rest
   C   least in the frame in which the centre of mass is at rest
   D   equal to the kinetic energy of the whole mass concentrated at the centre of mass.

5.7 Problems

5.1 Three particles of mass 1, 2 and 3 kg are located at $r_1 = \hat{i} + 2\hat{j} + 3\hat{k}$, $r_2 = 3\hat{i} + 2\hat{j} - \hat{k}$, $r_3 = 2\hat{i} - 3\hat{j} + \hat{k}$ respectively, where $\hat{i}, \hat{j}, \hat{k}$ are three mutually perpendicular unit vectors. Find the centre of mass of the system. (All distances are in m.)

5.2. The *centre of gravity* of a system of particles is defined as that
      point about which gravitational forces acting on the particles
      exert no net torque. Show that in a uniform gravitational field
      the centre of gravity of a system of particles coincides with the
      centre of mass.

5.3. Show that in an experiment in which protons bombard a target
      of deuterons, the CM energy is $\frac{2}{3}$ of the LAB energy.

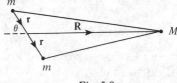

Fig. 5.8

*5.4 Show that the gravitational torque exerted by a mass M distance R
      from a gravitational dipole, consisting of two masses m, distance $2r$
      apart, where $r \ll R$, is approximately given by

$$N = -\frac{3GMmr^2}{R^3}\sin 2\theta$$

      where θ is the angle between **r** and **R**. Hence show that the angular
      acceleration of the dipole is

$$\ddot{\theta} = -\frac{3GM}{2R^3}\sin 2\theta.$$

*5.5 Supposing that, in problem 5.4, the two masses m represent the
      equatorial bulge of the earth, compare the torque exerted on the
      earth by the sun and the moon.

Solutions

1. (D) The centre of mass is at rest
or in uniform motion.
Note that (C) is quite mean-
ingless, since no frame of
reference is specified.

2. (C) This follows from eq. (5.11).

6. Solids

6.1 Introduction

So far we have been concerned with the motion of individual particles
or of extended bodies, which for the purpose under discussion could be
treated as individual particles. In this and in the following two chapters
we turn to the motion of matter in bulk. We know, of course, that
matter is made up of molecules, so that in principle we could describe
the motion of a piece of matter in terms of the motion of its individual
molecules, treated as point particles, but the number of molecules
contained in even the smallest piece of matter handled in everyday life
is so enormous—a grain of sand contains about 10^{15} molecules [see
section 8.5]—that such a treatment is impossible in practice. Instead,
we take advantage of the fact that the molecules are so small and
closely packed that matter appears to be continuous. We then con-
ventionally consider three states of matter:

(a) The solid state, in which matter keeps both shape and volume
constant.
(b) The liquid state, in which the volume but not the shape of a
given piece of matter remains constant.
(c) The gaseous state, in which neither volume nor shape are constant.

While the borders between these divisions are not entirely well defined,
so that for instance glass has been classified as both solid and liquid, the
division is useful, and in this chapter we shall deal with the motion of
solids. [See G. H. Wannier, 'The nature of solids', *SA* 249, December
1952, p. 39, and J. D. Bernal, 'The structure of liquids', *SA* 267,
August 1960, p. 124.]

6.2 The rigid body

Even for a solid, the statement that it keeps its shape is an over simplification, since it is possible to deform a solid, e.g., bend a steel bar. We
shall return to this point in section 6.8, but for the time being make the
assumption that we are dealing with a *rigid body*, in which such

deformations are ignored. If we go back to the description of matter in terms of molecules, then we can think of a rigid body as a collection of N particles of mass m_i at positions r_i, where $i = 1, 2, \ldots, N$, which is such that the distances

$$|\mathbf{r}_i - \mathbf{r}_j| = \text{constant for all } i, j. \tag{6.1}$$

These distances are maintained by means of balances between forces $\mathbf{F}_{ij}$ between the particles. A rigid body is therefore a special case of a system of particles.

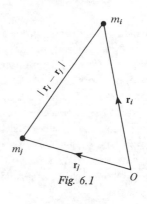

Fig. 6.1

6.3 Motion of a rigid body

It is a remarkable fact that, although the position of a single point can be fixed with the help of three quantities, such as, say, the co-ordinates in an orthogonal reference frame, the position of a rigid body, consisting of a very large number of such particles, needs only three more. To see this, we first show that the position of a rigid body is fixed, if the positions of any three of its points O, A, B, not lying in a straight line, are known. If the first of these is fixed, the body still can turn about it, as if there was a universal joint at this point. On fixing the second, we reduce the possible motion to a rotation about an axis through the two points, while fixing the third locks the body completely in position. Now the positions of three points can be described by means of nine co-ordinates, but because the body is rigid, the three distances between the points are constant. This leads to three relations between the nine co-ordinates, so that only six of them are independent. The number of independent quantities needed to describe the position of a mechanical system is called its *degrees of freedom*. Thus a point particle has three degrees of freedom and a rigid body six. The idea of degrees of freedom is particularly important in connection with the description of matter as made up of molecules, as we shall see in chapter 8.

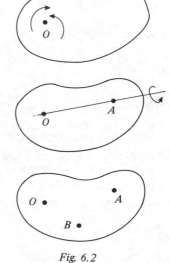

Fig. 6.2

It is clear that the motion of the body, i.e., its kinematics, is completely described by the motion of the triangle OAB. In considering two successive positions of this triangle, OAB and $O'A'B'$, we can imagine it to have moved first translationally and without rotation so that the point O moves to its new position O', while A and B move to new intermediate positions A'', B''. It can then be shown, although it is far from obvious [see worked example 6.1] that it is always possible to determine uniquely an axis $O'Z$ which is such that a single rotation about it through a definite angle will move A'' to A' and B'' to B'. Thus it is always possible to displace a rigid body from one position to another by a *translation* in which all particles of the body move through the same distance and in the same direction, say s, and a *rotation* through an angle θ about some axis, the direction of which is given by a unit vector n̂. These motions are independent of each other and can

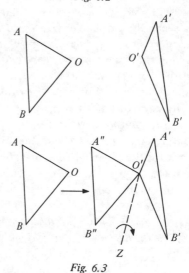

Fig. 6.3

be performed in any order, or even simultaneously. The choice of
point O is, of course, not unique, and so the magnitude and direction
of s would be different, if we chose a different point for O. On the other
hand, it can be shown that the angle θ and the direction of $\hat{n}$ do not
depend on the choice of O, so that the rotational motion is uniquely
determined by the displacement of the body as a whole.

Since the above arguments are true for any displacements, they
are certainly true for infinitesimal ones. If we divide the corresponding
translation ds and rotation $d\theta$ by the time dt taken over them, we
conclude that the motion of a rigid body can be described by a *velocity
of translation*

$$\mathbf{v} = \frac{d\mathbf{s}}{dt} \tag{6.2}$$

and an *angular velocity of rotation*

$$\boldsymbol{\omega} = \frac{d\theta}{dt}\hat{\mathbf{n}}, \tag{6.3}$$

where $\hat{n}$ is a unit vector along the axis of rotation. While $\mathbf{v}$ is not unique,
$\boldsymbol{\omega}$ is.

We shall be particularly concerned with two kinds of motion. In
the first, the rigid body is entirely free to move, as is for instance a
satellite in space. In that case it will be convenient to use the centre of
mass C of the rigid body for O, for then all the simplifying results of the
last chapter apply, and in particular the motion of C is given by (5.8),

$$\mathbf{F} = M\ddot{\mathbf{R}}, \tag{6.4}$$

where $\mathbf{F}$ is the resultant of the forces on the rigid body, and M the mass
of the rigid body. The rotational motion is then about an axis through C.

Secondly, we shall be concerned with the rotation of a rigid body
about an axis through one fixed point, such as a spinning top. Since we
showed in the last chapter that the motion of the centre of mass and that
about the centre of mass can be treated independently, it follows that the
free motion of a rigid body can be treated as the superposition of a
translational motion of the centre of mass and a rotational motion about
an axis through the centre of mass which for this part of the motion may
be considered as fixed. Thus for the rotational part of the motion, the
treatment is the same, whether one point of the body is fixed, or no
point is fixed, provided that in the latter case the axis of rotation is
taken to pass through the centre of mass.

*Worked example 6.1. Construct the axis of rotation for the rotational
part of the displacement of a rigid body.*

We replace the rigid body by a triangle OAB fixed in the body and
take A and B on a sphere, centre O. Then a rotation of the body about

O rotates the surface of the sphere over itself so that A and B move to A', B'. We now have to use geometry on the sphere, where straight lines are replaced by arcs of great circles. Draw the arcs AA' and BB' and let P and Q be the points bisecting AA' and BB' respectively. Let the great circle arc through P perpendicular to AA' meet the great circle arc through Q perpendicular to BB' in Z. Then OZ is the axis of rotation.

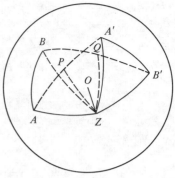

Proof. Because of the construction,

$$\widehat{AZ} = \widehat{A'Z} \quad \text{and} \quad \widehat{BZ} = \widehat{B'Z}.$$

Since A and B are fixed to the rotating sphere,

$$\widehat{AB} = \widehat{A'B'}.$$

Fig. 6.4

Hence the spherical triangles ABZ and $A'B'Z$ are congruent and the angles $\angle AZB$ and $\angle A'ZB'$ are the same. It follows at once that the rotation through angle $\angle AZA'$ which takes A to A' also takes B to B' since $\angle AZA' = \angle BZB'$ This completes the proof that the triangle OAB is rotated into triangle $OA'B'$ by a rotation about OX through an angle $\theta = \angle AZA'$.

6.4   Moment of inertia

Before dealing with the motion about an axis with one point fixed, we consider the very much simpler problem, when the axis has two points fixed, which means that it is completely fixed.

Let the external force on the particle m_i at $\mathbf{r}_i$ of the rigid body be $\mathbf{F}_i$. (We know from the last chapter that we do not need to consider internal forces.) If the body has an angular velocity ω about the axis of rotation Oz, and the distance of m_i from Oz is ρ_i, then the angular momentum of the body about Oz is

$$L_z = \sum m_i \rho_i^2 \omega = I_0 \omega, \tag{6.5}$$

where the quantity

$$I_0 = \sum m_i \rho_i^2 \tag{6.6}$$

is called the *moment of inertia* of the body about the axis Oz. In many ways, the moment of inertia plays the same role in rotational motion as does the mass in translational motion, but it should be noted that the moment of inertia depends not only on the mass, but also on how the mass is distributed over the volume of the body and also on the position of the axis relative to the mass distribution. Clearly, before we can make any progress with our investigation of rotational motion, we must learn more about how to calculate moments of inertia.

We first prove two theorems on moments of inertia of a very general nature, and then use these to calculate the moments of inertia of uniform bodies of simple geometrical shape.

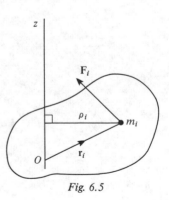

Fig. 6.5

Parallel axes theorem. This theorem enables us to find the moment of inertia about one axis, when that about a parallel axis through the centre of mass C is known. Clearly, if the distance between the axes is d,

$$I_O = \sum m_i \rho_i^2 = \sum m_i(\mathbf{d} + \boldsymbol{\rho}_i')^2$$

$$= \sum m_i(d^2 + 2\mathbf{d}\cdot\boldsymbol{\rho}_i' + \rho_i'^2)$$

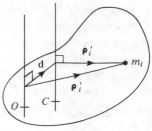

Fig. 6.6

Since $\sum m_i \boldsymbol{\rho}_i' = 0$ for distances from an axis through C, we have, if we choose one axis to pass through C,

$$I_O = I_C + Md^2. \tag{6.7}$$

Thus if we know I_C, we can calculate I_O.

Perpendicular axes theorem. This theorem applies only to a lamina, i.e., a rigid two-dimensional body. For this, the moment of inertia about an axis Oz perpendicular to the plane of the lamina, is

$$I_z = \sum m_i \rho_i^2 = \sum m_i(x_i^2 + y_i^2)$$

$$\therefore \quad I_z = I_y + I_x \tag{6.8}$$

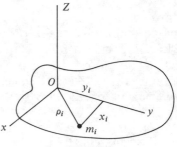

Fig. 6.7

Thus, if we know the moments of inertia about any two perpendicular axes in the plane of the lamina, we can find the moment of inertia about the axis perpendicular to the plane and through the same origin.

We can use these theorems to find the moments of inertia of simple bodies. In all cases the mass of the body is M.

(a) *A rod of length 2a, about a perpendicular axis through the centre.* Divide the rod into volume elements of length dx. The mass of such an element is

$$dM = M\frac{dx}{2a}.$$

Hence

$$I = \int_{-a}^{a} x^2\,dM = \int_{-a}^{a} x^2 M\frac{dx}{2a^2} = \tfrac{1}{3}Ma^2. \tag{6.9}$$

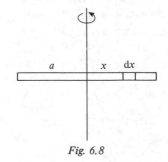

Fig. 6.8

(b) *A circular plate of radius a, about a perpendicular axis through the centre.* Divide the lamina into elements by means of concentric circles, distance dr apart. Then

$$dM = M\frac{2\pi r\,dr}{\pi a^2}$$

$$\therefore \quad I = \int_0^a r^2\,dM = \int_0^a \frac{2Mr^3\,dr}{a^2} = \tfrac{1}{2}Ma^2. \tag{6.10}$$

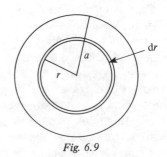

Fig. 6.9

(c) *A circular plate of radius a, about a diameter.* This must be the same for all diameters. If we consider two at right angles to each other, then it follows from (6.8) and (6.10) that

$$I = \tfrac{1}{4}Ma^2. \tag{6.11}$$

(d) *A rectangular plate with length of sides 2a and 2b about an axis parallel to one side.* We divided the plate into rod-shaped elements. Then

$$dM = M\,\frac{dx}{2b}$$

$$\therefore \quad I = \int_{-b}^{b} \tfrac{1}{3}M\,\frac{dx}{2b}\,a^2 = \tfrac{1}{3}Ma^2. \tag{6.12}$$

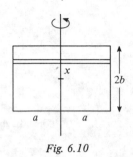

Fig. 6.10

(e) *A rectangular block with edges 2a, 2b, 2c, about an axis through the centre and parallel to one side.* We divided the block into a set of plates. For each of these the moment of inertia is found from (6.12) and (6.8),

$$dI = \tfrac{1}{3}dM(a^2 + b^2).$$

Hence for the block as a whole

$$I = \tfrac{1}{3}M(a^2 + b^2) \tag{6.13}$$

(f) *A sphere, of radius a, about a diameter.* The result (see problem 6.2) is

$$I = \tfrac{2}{5}Ma^2. \tag{6.14}$$

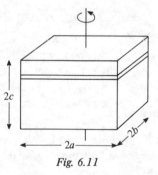

Fig. 6.11

Worked example 6.2. A pendulum consists of a rod of length 2a and mass m, to one end of which is attached a spherical bob of radius $\tfrac{1}{3}a$ and mass 15 m. Find the moment of inertia of the pendulum (a) about an axis through the other end of the rod and at right angles to the rod, (b) about a parallel axis through the centre of mass of the pendulum.

(a) We use the parallel axes theorem (6.7) separately on rod and sphere.

$$I_A(\text{rod}) = \tfrac{1}{3}ma^2 + ma^2 = \tfrac{4}{3}ma^2,$$

$$I_A(\text{sphere}) = \tfrac{2}{5}\cdot 15m\left(\frac{a}{3}\right)^2 + 15m\left(2a + \frac{a}{3}\right)^2$$

$$= \frac{247}{3}ma^2$$

$$\therefore \quad I_A = \frac{251}{3}ma^2.$$

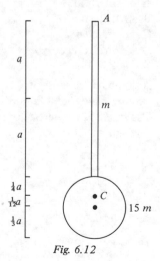

Fig. 6.12

(b) The centre of mass C is $\frac{1}{12}a$ from the centre of the sphere. (Check this!)

$$I_A(\text{rod}) = \tfrac{1}{3}ma^2 + m(\tfrac{5}{4}a)^2 = \tfrac{91}{48}ma^2,$$

$$I_A(\text{sphere}) = \tfrac{2}{5}\cdot 15m\left(\frac{a}{3}\right)^2 + 15m(\tfrac{1}{12}a)^2$$

$$= \tfrac{37}{48}ma^2$$

$$\therefore \quad I_C = \tfrac{8}{3}ma^2.$$

Note that $I_A - I_C = 81\,ma^2 = 16m(\tfrac{9}{4}a)^2$. This checks the parallel axes theorem for the whole pendulum.

Progress test

1. The moment of inertia of a rigid body of uniform density rotating about an axis does not depend on which one of the following?

   A   The total mass of the body
   B   The geometrical shape of the body
   C   The position of the axis of rotation
   D   The angular velocity of the body.

2. If the moment of inertia of a body rotating about an axis is doubled, but the angular momentum remains constant, then the angular velocity is

   A   quartered
   B   halved
   C   doubled
   D   quadrupled.

3. A dumbbell consists of a rod of length $2a$, mass $3M$, with a circular plate, radius a, mass $2M$, attached to each end as shown. The moment of inertia about an axis perpendicular to the rod and through its centre is

   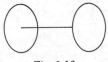

   Fig. 6.13

   A $\frac{4}{3}Ma^2$

   B $4Ma^2$

   C $6Ma^2$

   D $\frac{20}{3}Ma^2$.

4.   The moment of inertia of a body about an axis through its centre
     of mass is always

    A   less than that about any parallel axis
    B   equal to that about any parallel axis
    C   greater than that about any parallel axis
    D   none of these.

6.5   Rotation about a fixed axis

We now return to our investigation of rotational motion about a fixed
axis, i.e., to the angular momentum equation

$$L_z = I\omega, \tag{6.5}$$

where I is the moment of inertia about the axis Oz. The equation of
motion of the rotating body is given by the z-component of (5.25),

$$\dot{L}_z = I\dot{\omega} + \dot{I}\omega = N_z \tag{6.15}$$

where N_z is the z-component of the torque $\mathbf{N} = \Sigma\, \mathbf{r}_i \times \mathbf{F}_i$. In general, the
moment of inertia remains constant throughout the motion, which is
then given by

$$N_z = I\dot{\omega}, \tag{6.16}$$

but we shall note cases where this is not so. If there is no torque about
Oz, then L_z is constant, i.e., the angular momentum is conserved.

    The kinetic energy of rotation is

$$T = \Sigma\, \tfrac{1}{2}m_i v_i^2 = \Sigma\, \tfrac{1}{2}m_i \rho_i^2\, \omega^2$$

$$= \tfrac{1}{2}I\omega^2, \tag{6.17}$$

and if this energy results from the application of a constant torque N_z
rotating the body through a total angle θ in time t, then from (6.16)

$$N_z t = I\omega, \quad \tfrac{1}{2}N_z t^2 = I\theta. \tag{6.18}$$

Hence, work done

$$W = \tfrac{1}{2}I\omega^2 = N_z\, \theta. \tag{6.19}$$

    We may note the analogies between translational and rotational
motions, contained in (6.5), (6.15), (6.17), (6.19):

	Translational	Rotational
Momentum	$p = mv$	$L_z = I\omega$
Equation of motion	$F = m\dot{v} = \dot{p}$	$N_z = I\dot{\omega} = \dot{L}_z$
Kinetic energy	$T = \tfrac{1}{2}mv^2$	$T = \tfrac{1}{2}I\omega^2$
Work	$W = Fs$	$W = N_z\, \theta$

120 [6.5

Worked example 6.3. A flywheel, which may be taken to be a uniform circular disc, has a radius of 0·3 m and a mass of 20 kg. Calculate its angular momentum and energy, when it is rotating at 1200 revolutions per minute. What are the constant torque and average power needed to reach this angular velocity in 10 s?

The moment of inertia about an axis through the centre of and perpendicular to the flywheel is

$$I = \tfrac{1}{2} \times 20 \times (0\cdot3)^2 = 0\cdot9 \text{ kg m}^2.$$

The angular velocity is

$$\omega = \tfrac{1200}{60} \times 2\pi = 40\pi \text{ rad s}^{-1}.$$

$$\therefore \quad L = 36\pi \text{ m}^2 \text{ kg s}^{-1}.$$

$$T = 720\pi^2 \text{ J}.$$

$$\theta = \tfrac{1}{2}\omega t = 200\pi \text{ rad}.$$

$$N = T/\theta = 3\cdot6\pi \text{ Nm}.$$

Average power $= T/t = 72\pi^2$ W. Note that the maximum power is

$$N\omega = 144\pi^2 \text{ W}.$$

Worked example 6.4. A solid cylinder, a hollow cylinder and a sphere all roll without slipping down an inclined plane. If they are all released from rest at the same time from the same height, which will reach the bottom first?

There appears to be no fixed axis of rotation and so we analyse the motion into that of the centre of mass—motion in a straight line down the line of maximum slope—and motion relative to the centre of mass—rotation about a horizontal axis through C and parallel to the line of maximum slope. Relative to the centre of mass, the rotational axis is therefore fixed and we can use the simple theory of the present section.

As there is no slipping, the line of contact is at rest. Hence

$$\dot{s} = a\omega,$$

where $\dot{s}$ is the instantaneous velocity of C and ω the instantaneous angular velocity about C, when the body has descended a distance s. Energy conservation yields

$$\tfrac{1}{2}M\dot{s}^2 + \tfrac{1}{2}I\omega^2 - Mgs\sin\alpha = 0,$$

Solutions

1. (D) **Bookwork**

2. (B) $I_1\omega_1 = I_2\omega_2$

3. (C) $\tfrac{2}{5} \times 3Ma^2 + 2(\tfrac{1}{4} \times 2Ma^2 + 2Ma^2) = 6Ma^2$

4. (A) $I_c = I_A - Ma^2 < I_A.$

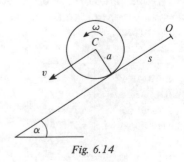

Fig. 6.14

or

$$\left(M + \frac{I}{a^2}\right)\dot{s}^2 = 2Mgs\sin\alpha,$$

where we have used (6.17) for the rotational kinetic energy. On differentiating, we obtain for the acceleration

$$\ddot{s} = \frac{Mg\sin\alpha}{M + I/a^2}.$$

Now the moments of inertia about the axis of rotation of the three bodies are

Solid cylinder: $I_C = \tfrac{1}{2}Ma^2$

Hollow cylinder: $I_H = Ma^2$

Sphere: $I_S = \tfrac{2}{3}Ma^2$

and the corresponding accelerations are

$$\ddot{s}_C = \tfrac{2}{3}g\sin\alpha, \qquad \ddot{s}_H = \tfrac{1}{2}g\sin\alpha, \qquad \ddot{s}_S = \tfrac{5}{7}g\sin\alpha.$$

Hence, the sphere reaches the bottom first, and this is so even when the masses and the radii of the three bodies are different from each other.

Progress test

1.   If the dumbbell in question 3 of the progress test for the previous section rotates at 10 rev s^{-1}, its kinetic energy is

   A $300Ma^2$

   B $600Ma^2$

   C $1200\pi^2 Ma^2$

   D $\dfrac{75}{\pi^2}Ma^2$

2.   The dumbbell is set rotating by applying a constant couple for two revolutions. The magnitude of the couple is

   A $150Ma^2$

   B $300\pi Ma^2$

   C $600\pi^2 Ma^2$

   D   none of these.

3. A cylinder rolls and slips on a horizontal plane. The friction between the cylinder and the plane is

    A   zero
    B   always backward
    C   always forward
    D   sometimes forward and sometimes backward.

4. A uniform rod, mass M and length $2a$, swings in a vertical plane about a horizontal axis through one end. If it starts from rest in the unstable vertical position, its angular velocity, as it passes through the lowest position, is

    A $\sqrt{g/a}$

    B $\sqrt{3g/4a}$

    C $\sqrt{3g/a}$

    D $\sqrt{3Mg/a}$.

5. At the same moment the angular acceleration is
    A g/a

    B $3g/4a$

    C $3g/a$

    D   none of these.

6.6   General motion of a rigid body

The general treatment of the motion of a rigid body free from con-straints is beyond the scope of this book, and we shall confine ourselves to a largely qualitative discussion of the case of a body that has an axis of symmetry about which it is rotating.

Consider such a body moving freely in space under no external forces, in the direction of its axis of symmetry. If it does not rotate about this axis, then even the slightest transient torque N, acting for a short time t about an axis at right angles to the axis of symmetry, will give the body a permanent angular velocity about the axis of the torque. This leads to indefinitely continuing rotation, so that the symmetry axis changes direction in space. On the other hand, if the body is initially rotating rapidly about the axis of symmetry and therefore has a large angular momentum about this axis, then the angular momentum due to the transient torque will add vectorially to it. This results in the symmetry axis deviating very slightly from its initial direction, but this deviation will stop when the torque stops, since a continuous torque is

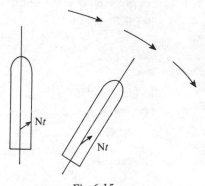

Fig. 6.15

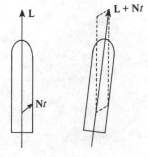

Fig. 6.16

needed to cause a continuous change in the angular momentum vector.
The stability induced by the spin about the symmetry axis is called the
gyroscopic effect, since it forms the basis of the working of the gyroscope.
The same effect is used in rifles, where the barrel has a helical groove cut
into it. This sets the bullet spinning along its axis, which ensures that it
continues to point in the direction of its motion after leaving the barrel.

The gyroscope is a device which consists of a heavy rotating flywheel, which is mounted in such a manner that its axis can freely
change direction. This can be achieved by supporting it on a universal
joint, or, more usually, in what is known as a *gimbal mounting*. This
consists of an outer and an inner ring. The outer ring turns freely about
a vertical axis fixed to an external support, while the inner ring turns
freely about a horizontal axis fixed to the outer ring. The flywheel
rotates about an axis fixed to the inner ring, which is at right angles
to both the other axes. As a result of this arrangement, any torque on
the external support does not transfer itself to the flywheel, which
continues to point in the same direction in space. Further, if there is a
little friction in the bearings, which transfers part of the torque, the
gyro effect mentioned above counteracts this. For this reason, this
arrangement is used in inertial guiding systems in ships and aeroplanes.
The flywheel is generally kept in motion by means of a turbine incorporated in it.

In the *gyrocompass* the inner and outer rings are fixed to each
other, and the external casing is arranged to move freely in a
horizontal plane, say by floating it on mercury.

The axis of rotation of the earth makes an angle with the vertical
(except at the poles, where the gyrocompass does not work), and the
resulting horizontal component L_H of the angular momentum $\mathbf{L}_E$ of
the earth leads to a torque on the outer ring about a north–south axis
which is transmitted to the flywheel, whose angular momentum is $\mathbf{L}$.
Just as in Fig. 6.16, this turns the axis of the flywheel towards the axis
of the torque, i.e., north–south. Thus the system acts as a compass.

Let us investigate more generally what happens when a torque is
applied to an axially symmetric body, rotating about its axis of symmetry
with angular momentum $\mathbf{L}$. Consider a torque

$$\mathbf{N} = \mathbf{b} \times \mathbf{F} \tag{6.20}$$

applied perpendicular to the axis, say on the bearings. As $\mathbf{N}$ is perpendicular to $\mathbf{L}$, it cannot change its magnitude. On the other hand, since
$\mathbf{N} = \dot{\mathbf{L}}$, it must change its direction. As $\mathbf{N}$ points into the paper, so does
$\mathbf{L}$, i.e., the upper bearing moves into the paper and the lower out of it.
This rotation of the symmetry axis continues as long as $\mathbf{N}$ acts. It is
called *precession*, and is a rotation of the symmetry axis about an axis
perpendicular to both $\mathbf{L}$ and $\mathbf{N}$. In fact, the precessional angular
momentum $\mathbf{P}$, together with $\mathbf{L}$ and $\mathbf{N}$, form a right-handed set of
orthogonal vectors. [Fig. 6.20.]

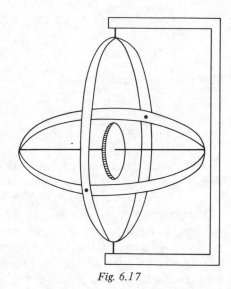

Fig. 6.17

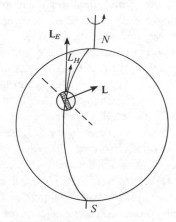

Fig. 6.18

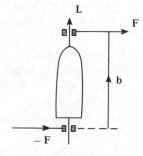

Fig. 6.19

To derive the rate of precession, we note that in a small time dt, L turns through an angle dφ, where

$$dL = L\, d\varphi. \tag{6.21}$$

Thus the rate of precession is

$$\dot\varphi = \frac{\dot L}{L} = \frac{N}{L} = \frac{Fb}{I\omega}, \tag{6.22}$$

where ω is the angular velocity and I the angular momentum of the body about the symmetry axis. Note that this equation is valid only if $\dot\varphi \ll \omega$, which is usually the case, for we have assumed that the total angular momentum vector acts along the symmetry axis. This is strictly true only at the beginning of the precessional motion, when $\varphi = 0$, but remains approximately true as long as $\dot\varphi \ll \omega$.

A well-known illustration of precession is the motion of a spinning top. The top has an angular momentum **L** about its symmetry axis which is inclined at some angle to the vertical. It suffers a torque **N** due to its weight and the normal reaction at the point of support. This leads to a change d**L** in time dt as shown. In consequence, the symmetry axis precesses about the vertical axis.

An application of the spinning top on the grand scale occurs in the motion of the earth. Here the torque is due to the attraction of the sun S and moon M on the equatorial bulge [see problem 5.5]. This tries to turn the equatorial plane into the plane of the ecliptic (the plane of the earth's orbit) and as a result the earth's axis precesses in a cone about a normal to the ecliptic, with a period of about 27 000 years. This is known as the *precession of the equinoxes*.

At the opposite extreme, we have the precession of electrons in a magnetic field. Electrons have spin angular momentum and a magnetic moment. [See Jackson, chapter 5]. This causes a torque in a constant magnetic field and a consequent precession with a frequency dependent on the strength of the field. This frequency can be detected by superimposing on the constant field a small magnetic field of alternating frequency. When this latter frequency is equal to the precession frequency, a resonance occurs and energy is transferred from the field to the electrons. This phenomenon of *electron spin resonance* and the similar one of *nuclear spin resonance* is of great importance in the study of the properties of atoms and nuclei.

A final application of the theory of precession occurs in the motion of a bicycle. The stability of a bicycle is largely due to the rider continually checking his incipient fall, i.e., use of the centrifugal force – in the frame of reference of the rider – when steering into a fall. However, the gyroscopic effect on the rapidly spinning wheels also plays a part and precession is important when a turn is made. Thus

Solutions

1. (C) $\omega = 10 \times 2\pi$ rad s^{-1},
$T = \frac{1}{2}I\omega^2$.

2. (B) $1200\pi^2 Ma^2 = 2 \times 2\pi \times N$

3. (D)   It is forward if $v < a\omega$,
backward if $v > a\omega$.

4. (C) $\frac{1}{2}I\omega^2 = 2Mga$, $I = \frac{4}{3}Ma^2$.

5. (D)   It is zero. $I\ddot\theta = L = N$ and the weight of the rod has no moment about the point of suspension in this position.

Fig. 6.20

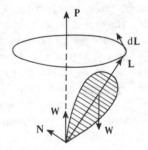

Fig. 6.21

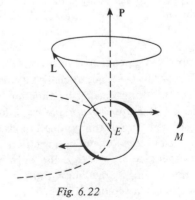

Fig. 6.22

when turning left, one should lean to the left and let the handlebars be turned through precession. One should not turn the handlebars to the left, for then the whole bicycle will precess to the right with disastrous consequences. The diagrams are viewed from the rider towards the handlebars. [For a full investigation of the problem, including a magnificent series of experiments, see D. E. H. Jones, 'The Stability of the Bicycle', *Physics Today*, April 1970.]

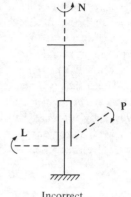

Incorrect

Progress test

1.   A gyrocompass which points north in England (longitude 0°, latitude 50°N) will point south in

    A   longitude 0°, latitude 50°S
    B   longitude 180°, latitude 50°N
    C   longitude 180°, latitude 50°S
    D   none of these.

2.   If a spinning body is to precess, it is essential that the torque on the body

    A   is very large
    B   is very small
    C   has a component parallel to the angular momentum vector
    D   has a component perpendicular to the angular momentum vector.

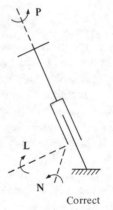

Correct

Fig. 6.23

3.   A person in a swivel chair holds a flywheel with its axle horizontal and rotating as shown. In order to turn himself to the right, he must, while holding his right hand steady, move the left hand

    A   up
    B   down
    C   inwards
    D   outwards.

6.7   Rotating frames of reference

At the end of the last section, and also briefly in chapter 2, there occurred a reference to centrifugal forces, which are experienced in a non-inertial rotating frame of reference. Apart from this, we have limited our discussion so far to inertial frames. However, we know that the earth is, strictly speaking, not an inertial frame, since it rotates about its axis. Let us consider the events referred to a rotating frame.

    There is a well-known experiment in which a person standing on a rotating turntable with weights in his outstretched arms is asked to bring his hands suddenly to his shoulders. As a result, he suddenly speeds up. (If the person is a girl, this is referred to as the experiment

Fig. 6.24

of the three dumb belles.) Further, as his arms are brought in, he notices the horizontal pull of the weights, which he describes as due to the centrifugal force. We, standing on the firm floor, can see no agent that produces such a force and call it an inertial, i.e., an 'unreal' force. (See section 2.3.)

There are no forces with a turning moment about the person, as his arms are brought in, and so his angular momentum remains unchanged. However, by bringing the weights nearer to his axis of rotation, he has decreased his moment of inertia. Hence his angular velocity must increase. Denoting before and after the change by suffixes 1, 2, we have

$$I_1 \omega_1 = I_2 \omega_2 \quad \text{and as } I_1 > I_2, \quad \omega_1 < \omega_2. \tag{6.23}$$

The total kinetic energy is given by

$$T_1 = \tfrac{1}{2} I_1 \omega_1^2, \ T_2 = \tfrac{1}{2} I_2 \omega_2^2 = \tfrac{1}{2} I_1 \omega_1 \omega_2 > T_1. \tag{6.24}$$

The increase in the kinetic energy is, in fact, equal to the work done against the centrifugal force! [See worked example 6.5.]

Now consider the body separately from the arms. The body has not changed its moment of inertia and yet it is rotating faster. There must therefore have been a torque acting on it, and this cannot have been due to the centrifugal force, since this is radial through the axis of rotation. It must be due to a couple consisting of horizontal forces at right angles to the centrifugal forces, acting in a forward and backward direction on the shoulder joints. It is called the *Coriolis force* after its discoverer, and occurs whenever a body moves in a rotating reference frame. Like the centrifugal force it is inertial, but to someone living in a rotating reference frame it is very real.

We next show that all these observations follow from an analysis of motion in a rotating reference frame $O'x'y'z'$. Let this frame have a common z-axis with an inertial frame $Oxyz$, and let it rotate with constant angular velocity ω about Oz, relative to the inertial frame. Consider the motion of a point in the xy-plane and let its position vector in the rotating frame be $\mathbf{r}$. (We shall confine ourselves to motion in the xy-plane throughout this discussion.) After time dt, $\mathbf{r}$ has undergone two changes, as viewed from the inertial frame:

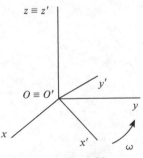

Fig. 6.25

(1) d$_1\mathbf{r}$, due to a change of $\mathbf{r}$ in the rotating frame.
(2) d$_2\mathbf{r}$, due to the rotation of the frame.

Clearly $d_2\mathbf{r}$ is at right angles to both $\boldsymbol{\omega}$ and $\mathbf{r}$ and of magnitude $\omega r \, \mathrm{d}t$. Hence from Fig. 6.26,

$$\mathbf{d_2 r} = \boldsymbol{\omega} \times \mathbf{r} \, \mathrm{d}t \tag{6.25}$$

and

$$\mathbf{dr} = \mathbf{d_1 r} + \boldsymbol{\omega} \times \mathbf{r} \, \mathrm{d}t. \tag{6.26}$$

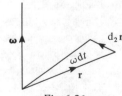

Fig. 6.26

Hence the rates of change of $\mathbf{r}$ in the frames $Oxyz$ and $O'x'y'z'$ are related through the equation

$$\left(\frac{d\mathbf{r}}{dt}\right)_O = \left(\frac{d\mathbf{r}}{dt}\right)_{O'} + \boldsymbol{\omega} \times \mathbf{r} \tag{6.27}$$

or

$$\mathbf{v} = \mathbf{v}' + \boldsymbol{\omega} \times \mathbf{r}. \tag{6.28}$$

It can be shown that the relation (6.27), which relates the rates of change of the vector $\mathbf{r}$ in the two frames, is valid for the relation of the rates of change of any vector in the two frames. Applying it in turn to $\mathbf{v}$ and $\mathbf{v}'$, we have

$$\left(\frac{d\mathbf{v}}{dt}\right)_O = \left(\frac{d\mathbf{v}}{dt}\right)_{O'} + \boldsymbol{\omega} \times \mathbf{v} \tag{6.29}$$

and

$$\left(\frac{d\mathbf{v}'}{dt}\right)_O = \left(\frac{d\mathbf{v}'}{dt}\right)_{O'} + \boldsymbol{\omega} \times \mathbf{v}'. \tag{6.30}$$

Now

$$\left(\frac{d\mathbf{v}}{dt}\right)_{O'} = \left[\frac{d}{dt}\left(\frac{d\mathbf{r}}{dt}\right)_O\right]_{O'}$$

$$= \left[\frac{d}{dt}\left(\frac{d\mathbf{r}}{dt}\right)_{O'}\right]_O = \left(\frac{d\mathbf{v}'}{dt}\right)_O, \tag{6.31}$$

since the order of differentiation can always be interchanged. Hence

$$\left(\frac{d\mathbf{v}}{dt}\right)_O = \left(\frac{d\mathbf{v}'}{dt}\right)_{O'} + \boldsymbol{\omega} \times \mathbf{v}' + \boldsymbol{\omega} \times \mathbf{v}$$

or, using (6.28) and replacing $d\mathbf{v}/dt$ by $\mathbf{f}$,

$$\mathbf{f} = \mathbf{f}' + 2\boldsymbol{\omega} \times \mathbf{v}' + \boldsymbol{\omega} \times (\boldsymbol{\omega} \times \mathbf{r}). \tag{6.32}$$

Now $\boldsymbol{\omega}$ is perpendicular to $\mathbf{r}$, and hence $\boldsymbol{\omega} \times \mathbf{r}$ is perpendicular to both and of magnitude ωr. Next, $\boldsymbol{\omega} \times (\boldsymbol{\omega} \times \mathbf{r})$ is perpendicular to $\boldsymbol{\omega}$ and $\boldsymbol{\omega} \times \mathbf{r}$ and thus in the direction of $-\mathbf{r}$ and of magnitude $\omega^2 r$. Hence

$$\boldsymbol{\omega} \times (\boldsymbol{\omega} \times \mathbf{r}) = -\omega^2 \mathbf{r}, \tag{6.33}$$

so that

$$\mathbf{f} = \mathbf{f}' + 2\boldsymbol{\omega} \times \mathbf{v}' - \omega^2 \mathbf{r}. \tag{6.34}$$

We therefore conclude, that for a particle of mass m we have the relation

$$m\mathbf{f}' = m\mathbf{f} + m\omega^2 \mathbf{r} - 2m\boldsymbol{\omega} \times \mathbf{v}'. \tag{6.35}$$

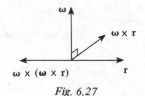

Fig. 6.27

Solutions

1. (D) A gyrocompass points in the same direction every-where. (Draw Fig. 6.18 for a point in the southern hemisphere.)

2. (D) Bookwork.

3. (A) He must turn the axle clockwise, as seen from himself, since $\mathbf{L}$ is to the left and $\mathbf{P}$ down.

The left-hand side is what an observer in the rotating frame measures as the total force on the particle. It is made up of three parts:

$m\mathbf{f}$ —the 'real' force

$m\omega^2\,\mathbf{r}$—the centrifugal force, outwards along the radius

$-2m\boldsymbol{\omega}\times\mathbf{v}'$—the Coriolis force.

It will be seen that the Coriolis force has exactly the properties that we deduced for it from our experiment at the beginning of this section:

(a)   It is at right angles to the axis of rotation and to the velocity of the particle in the rotating frame.

(b)   It vanishes when the velocity of the particle is zero.

The Coriolis force must be allowed for in long-range gunnery and of course in missiles. Another important application concerns the overall pattern of the earth's winds. Hot air rises all along the equator, and gets displaced into higher and cooler latitudes, where it sinks down to ground level. It then spreads north and south, and as is easily seen, is influenced by the Coriolis force so that the streams that spread away from the equator are deflected to the east (westerlies and roaring forties), while those that spread towards the equator are deflected to the west (north-east and south-east trade winds).

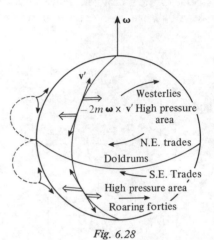

Fig. 6.28

Worked example 6.5. Show that the increase in kinetic energy in the 'dumb belle' experiment is equal to the work done against the centrifugal force.

Let the moment of inertia of the person be I_0, and let the distance of the weights, of combined mass M, from the axis of rotation be r at any given time. For the sake of simplicity we ignore the weight of the person's arms. Initial and final positions are denoted by 1 and 2.

The angular momentum, which is constant, is given by

$$L = (I_0 + Mr^2)\,\omega$$

and the centrifugal force, in the rotating frame of reference, is

$$F = Mr\omega^2.$$

Hence the work done against the centrifugal force is (the $-$ sign allows for the 'against')

$$W = -\int_{r_1}^{r_2} Mr\omega^2\,\mathrm{d}r$$

$$= -\int_{r_1}^{r_2} \frac{ML^2 r\,\mathrm{d}r}{(I_0 + Mr^2)^2}$$

$$= \left[\frac{L^2}{2(I_0 + Mr^2)} \right]_{r_1}^{r_2}$$

$$= [\tfrac{1}{2}(I_0 + Mr^2)\omega^2]_1^2$$

$$= T_2 - T_1.$$

Progress test

1.   If a person is spinning with outstretched arms and suddenly brings his arms down to his side, his kinetic energy

    A   increases
    B   decreases
    C   first increases and then decreases to what it was before
    D   remains unchanged.

2.   The Coriolis force on a body in a rotating frame of reference does not depend on which one of the following?

    A   The mass of the body
    B   The position of the body referred to the frame
    C   The velocity of the body referred to the frame
    D   The angular velocity of the rotating frame referred to an inertial frame.

6.8   Deformation in solids

So far we have considered solid bodies to be rigid, but in fact all solids can be deformed to a greater or lesser degree. The extent of deformation for a given applied force depends not only on the shape of the body, but also on the material of which the body is constructed. For a full account of the effects of deformation it is therefore necessary to turn to books dealing with the science of materials [see, e.g., M. T. Sprackling, *The Mechanical Properties of Matter*, English Universities Press, 1970], and we shall here confine ourselves to a few general principles.

If we consider a bar under tension, then the resulting deformation will depend not only on the applied tension force F but also on the cross-sectional area A of the bar. It is intuitively obvious that a true measure of the agent of deformation is the ratio of force to area, since we can think of the bar as made up of a number of parallel bars of smaller cross-sectional area. This ratio is called the *stress* and is defined as

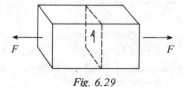

Fig. 6.29

$$\text{stress} = \frac{\text{force}}{\text{area}} = \frac{F}{A}. \tag{6.36}$$

There are different ways of applying stress to a body, and the one shown in Fig. 6.29 is called *tensile stress*. This has the effect of elongating the

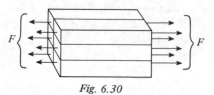

Fig. 6.30

body. *Compressive stress* has the opposite effect to tensile stress, while another, which deforms the shape of the body is called *shear stress*. [See Fig. 6.31.] We shall confine ourselves to the effect of tensile stress.

The effect of a stress is to produce a *strain*, which is defined as the proportional change due to the deformation. Thus for a material of length l which is extended by length Δl,

$$\text{tensile strain} = \frac{\text{extension}}{\text{original length}} = \frac{\Delta l}{l}. \tag{6.37}$$

Thus, while the dimensions of stress are $ML^{-1}T^{-2}$, strain is a dimensionless quantity.

The relation between stress and strain can be very different for different materials. For small stresses, most materials are *elastic*, i.e., they return to their original shape when the stress is removed. Beyond this stress, brittle materials tend to crack, while others, such as most metals, become *plastic*. If a material is stressed into the plastic region, then removal of the stress will not lead to a return to the original shape. Instead, a permanent deformation, known as a *permanent set*, will remain. If the stress is increased further, then eventually the cross-sectional area of the material is reduced to such an extent that elongation occurs even at reduced stress, i.e., the material *runs away*. Beyond this, *fracture* occurs. All this is shown diagrammatically in Fig. 6.32, where OB is the elastic and BD the plastic region, while fracture occurs at E. The stress at B is called the *yield stress* and that at E the *breaking stress*. It should be noted that the stress plotted is the *engineering stress*, defined as force/original cross-sectional area, which is the quantity of importance to the design engineer. The true or *physical stress* defined as force/actual cross-sectional area is larger, since the deformation due to the stress decreases the cross-sectional area, and it increases up to the breaking point E. If the stress is removed at, say, C in the plastic region, then the material returns to F, where OF is the permanent set. For small stresses, the stress–strain curve is linear. This region, i.e., OA, in which stress is proportional to strain, is called the *Hooke region*, since Robert Hooke (1635–1703) was the first to enunciate the law of proportionality of stress and strain. An example of Hookean behaviour occurred in the extension of a spring in worked example 1.7.

The slope of the stress–strain curve in the Hooke region is called the *elastic modulus*. Different moduli apply to the different types of stress referred to, and for tensile and compressive stress we have *Young's modulus*

$$Y = \frac{\text{tensile stress}}{\text{tensile strain}} = \frac{\text{compressive stress}}{\text{compressive strain}} \tag{6.38}$$

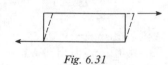

Fig. 6.31

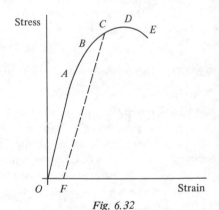

Fig. 6.32

or

$$Y = \frac{F/A}{\Delta l/l}.$$ (6.38')

As strain is a pure number, the dimensions of Y are the same as those of stress, and Y is measured in Nm^{-2}.

A true understanding of the relationship between stress and strain in a given material can be obtained only through an investigation of the molecular structure of the material. For this, the reader is referred to Sprackling's book.

Worked example 6.6. A copper wire of length 10 m has a modulus of elasticity, Y = 11·0 × 10¹⁰ Nm⁻², a yield stress of 3·3 × 10⁸ Nm⁻² an engineering breaking stress of 3·0 × 10⁸ Nm⁻² and a reduction of area at breaking point of 75 per cent. Calculate (a) the diameter of the wire, assuming it can just support a load of 250 kgf, (b) the extension under this load, (c) the physical stress at breaking point.

(a) $\frac{1}{4}\pi d^2 \times 3\cdot3 \times 10^8 = 250$ g

$$d = \sqrt{\frac{1000 \times 9\cdot8}{\pi \times 3\cdot3 \times 10^8}}$$

$$= 0\cdot0030 \text{ m} = 3\cdot0 \text{ mm}.$$

(b) $\dfrac{\Delta l}{l} = \dfrac{3\cdot3 \times 10^8}{11\cdot0 \times 10^{11}} = 3 \times 10^{-4}$

$$= 0\cdot03 \text{ per cent}$$

$$\Delta l = 3 \text{ mm}.$$

(c) Physical stress $= 3\cdot0 \times 10^8/0\cdot25$

$$= 12 \times 10^8 \text{ Nm}^{-2}.$$

6.9 Problems

6.1. Find the number of degrees of freedom of

   (a)   a small bead threaded on a thin fixed wire,
   (b)   a thin rod,
   (c)   a thin triangular lamina.

6.2. By dividing a sphere into thin circular parallel plates, prove (6.14).

6.3. A cylinder of mass M, radius r, is set to rotate with angular velocity ω_0 about its own axis, which is fixed. If after time t the angular velocity is ω_1, find the frictional torque N on the cylinder, assuming that

(a) N is constant
(b) $N = k\omega$, where ω is the instantaneous angular velocity.

*6.4. A rigid body is acted on by a set of forces $\mathbf{F}_i$. Show that these can be reduced to a suitably chosen single force $\mathbf{F}$ and couple $\mathbf{G}$.

6.5. A billiard ball is struck by a cue moving horizontally in the vertical plane through the centre of the ball. (a) At what height must it be struck if there is to be pure rolling? (b) Show that, if the ball is struck at a higher point, the velocity of the ball at first increases. [The dynamics of billiards are discussed in some detail in A. Sommerfeld, *Mechanics*, Academic Press, 1964.]

6.6 Starting from rest at the top, a particle slides down a large sphere with a frictionless surface. (a) At what point will the particle leave the sphere? (b) In what way would the answer differ, if the particle were replaced by a small sphere that rolled without slipping down the large sphere?

6.7 Two equal flywheels A and B are mounted freely on a shaft and may be connected or disconnected through a clutch C. At first, A is spinning with angular velocity ω_0 and B is at rest. They are then connected through the clutch. If in the process the clutch develops a quantity of heat H, what was the original kinetic energy of A?

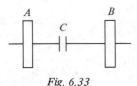

Fig. 6.33

6.8. A solid sphere spins about a diameter as an axis, and is free to pivot about a point on the axis at the surface of the sphere. What is the precessional angular velocity if the rotational angular velocity of the sphere is ω and the axis is horizontal?

6.9. In the 'dumb belle' experiment, let the angular velocity ω be maintained constant and consider the rate of change of angular momentum of the weights, each of mass m, as they are being drawn in. Hence show that the Coriolis force on each weight is given by

$$F = 2\,m\omega v$$

at right angles to the radius, where v is the velocity with which the weights are being drawn in.

6.10. There is another, apparently more intuitive, explanation of the wind pattern on the Earth, than the one given in section 6.7. According to this explanation, air moving north tries to maintain its speed constant along a parallel of latitude, while the speed of the earth's surface in the same direction decreases as the distance from the earth's axis decreases. Is this a correct explanation and, if so, how is it related to the explanation in section 6.7?

6.11. A certain alloy has a modulus of elasticity of 18×10^{10} Nm^{-2}, and a yield stress of 3.4×10^8 Nm^{-2}.

    (a)   What load could be supported by a wire of length 8·0 m and 4·0 mm diameter?

    (b)   What is the maximum permissible load if the extension of the wire must not exceed 2·0 mm?

6.12. In accounts of the Bohr model of the hydrogen atom, in which a negatively charged electron moves in a circle round a positively charged nucleus, one textbook states that 'the centripetal force is provided by the electrostatic force' [A. Beiser, Concepts of Modern Physics, McGraw-Hill, 1967] and another that 'the centrifugal force balances the electrostatic attraction' [Jackson, chapter 4]. Who is right or are both right?

7. Fluids

7.1 Ideal fluids

In the last section of chapter 6 we considered the elastic properties of a solid body. These were such that restoring forces were set up, when the shape of the body was changed, even though the volume, by and large, was not. The essential property of a fluid is that there are no such restoring forces for a change of shape without a change of volume. To the extent that this is only approximately true for real fluids, we shall confine courselves to *ideal fluids* for which this condition is exactly satisfied. It follows that there are no shearing forces in ideal fluids. Hence, if we draw a small imaginary surface dS inside the fluid, then the force dF on this surface due to the fluid, which is called the *thrust*, must be normal to the surface. The thrust per unit area is called the *pressure* on the surface, that is,

$$p = \frac{\mathrm{d}F}{\mathrm{d}S}.$$ (7.1)

It is closely related to compressive stress, as defined in section 6.8.

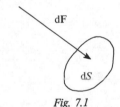

Fig. 7.1

7.2 Pressure in a fluid

In order to investigate the pressure in a fluid, we consider a small volume element in the shape of a tetrahedron, with three of its sides parallel to the co-ordinate planes and the fourth side at an arbitrary angle to these planes. Let the pressures on the four sides be p_x, p_y, p_z and p, as shown, and let the resultant external force on the mass of fluid contained in the volume element be **F**. This external force may be due to gravity or—in an accelerated motion of the fluid—may be due to inertial forces, but it is, in any case, proportional to the volume of the element. The thrusts on the sides of the volume element, on the other hand, are proportional to the areas of these sides. Hence in the limit, as the volume element shrinks to zero, the effect of the external forces is vanishingly small, compared to that of the thrusts.

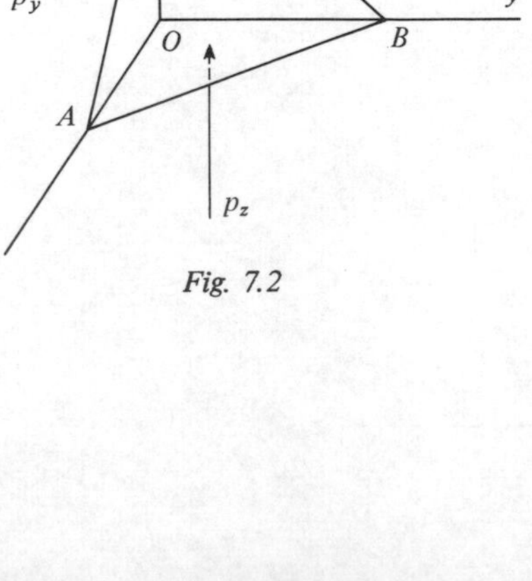

Fig. 7.2

When considering the equilibrium of the volume element under the forces acting on it, we may therefore neglect the external forces. Now, if the angle between the triangles ABC and OBC is θ, then

$$\Delta OBC = \Delta ABC \cos \theta. \tag{7.2}$$

Taking components of the forces on the element along the x-axis, we have

$$p\Delta ABC \cos \theta = p_x \, \Delta OBC,$$

and hence,

$$p = p_x. \tag{7.3}$$

Similarly, $p = p_y = p_z$. Thus, the pressure on our element of surface at a given point is independent of the direction of the surface element, and we therefore speak of *pressure at a point* of a fluid. Alternatively, we may say that the normal pressure on a volume element of fluid is the same in all directions.

We next relate the pressure to the density of the fluid at the point. Here we distinguish two limiting cases. The first is the *ideal liquid*, which is totally incompressible, so that its density is constant throughout. The second is the *ideal gas*, for which pressure and density at a given temperature are related through Boyle's law which, in one of its forms, states that the two are proportional to each other. For a fuller discussion of Boyle's law and the concept of temperature, see chapter 8.

To illustrate the relation between pressure and density in a particular case, we consider the variation of pressure in a column of fluid of density ρ under gravity. We take the z-direction along the constant gravitational field, and consider the equilibrium of a small cylindrical element of length $\mathrm{d}z$ and cross-sectional area $\mathrm{d}A$ under the external force and the opposing thrusts on the two ends of the cylinder:

$$p \, \mathrm{d}A - (p + \mathrm{d}p) \, \mathrm{d}A - \rho g \, \mathrm{d}A \, \mathrm{d}z = 0.$$

Hence,

$$\frac{\mathrm{d}p}{\mathrm{d}z} = -\rho g. \tag{7.4}$$

For an ideal liquid, ρ is constant, and hence we obtain

$$p = p_0 - \rho g z, \tag{7.5}$$

where p_0 is the pressure at the level $z = 0$. It is usual to take this as the surface of the liquid, where the pressure due to the liquid vanishes, so that (7.5) reduces to

$$p = -\rho g z, \tag{7.5'}$$

i.e., the pressure in a liquid under gravity is proportional to the depth below the surface. On the other hand, for an ideal gas at constant

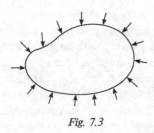

Fig. 7.3

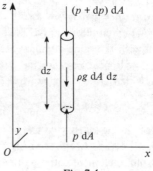

Fig. 7.4

temperature, we have

$$\frac{p}{\rho} = \frac{p_0}{\rho_0},$$

(7.6)

where p_0 and ρ_0 are the pressure and density at $z = 0$. Substituting in (7.4), we have

$$\frac{\mathrm{d}p}{\mathrm{d}z} = -\frac{\rho_0}{p_0}\, gp,$$

and hence,

$$p = p_0\, e^{-(\rho_0/p_0)gz}.$$

(7.7)

This is known as the pressure law for an isothermal atmosphere, i.e., the pressure in an ideal gas at constant temperature under gravity decreases exponentially with height. For small height differences, the assumption of constant temperature is reasonable, and enables us to use a barometer to measure height above ground level, e.g., when climbing a hill.

Surfaces which join points at which the pressure is the same are called *surfaces of equipressure*. In the above case, these are clearly planes at right angles to the gravitational field. For a liquid in a container, the free surface, which is in contact with the atmosphere, must always be a surface of equipressure. In the free surface, the pressure due to the liquid is zero, and the total pressure is therefore equal to the atmospheric pressure.

Worked example 7.1. Determine the free surface of a liquid in equilibrium in a container which is rotating with constant angular velocity.

When equilibrium has been reached, the liquid and container rotate with the same angular velocity, say ω. Taking the z-axis vertically, and co-ordinate axes rotating with the liquid, the external force on a volume element $\mathrm{d}V$ is given by (6.35),

$$\mathbf{F} = \rho\,\mathrm{d}V(\mathbf{g} + \omega^2\,\mathbf{r} - 2\boldsymbol{\omega} \times \mathbf{v}')$$

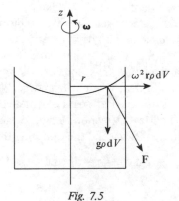

Fig. 7.5

where ρ is the density of the liquid and $\mathbf{v}'$ the velocity of the volume element relative to the rotating frame of reference. Hence $\mathbf{v}' = 0$. Now, the volume element must be in equilibrium under this force and the force resulting from the pressure. Hence the surfaces of equipressure must be at right angles to $\mathbf{F}$, as is easily seen from the figure. Hence, the angle θ between the free surface and the horizontal plane is given by

$$\tan\theta = \frac{\omega^2 r}{g}$$

that is,

$$\frac{\mathrm{d}z}{\mathrm{d}r} = \frac{\omega^2 r}{g},$$

Surface of equipressure

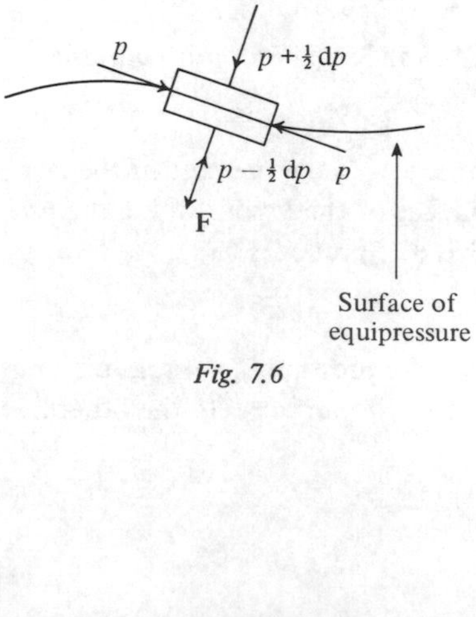

Fig. 7.6

or

$$z = z_0 + \frac{\omega^2}{2g} r^2,$$

so that the free surface is parabolic.

Progress test

1.   An ideal liquid is

     A   rigid
     B   compressible
     C   deformable
     D   none of the above.

2.   The surfaces of equipressure for a fluid rotating in free space under its own gravitational attraction are

     A   plane
     B   parabolic
     C   spherical
     D   none of the above.

7.3   Equilibrium of floating bodies

If we consider a fluid in static equilibrium and describe a closed surface inside this fluid, then the resultant force due to the fluid pressure on this surface must be equal and opposite to the resultant of the external forces acting on the fluid inside the surface. In particular, if the fluid is under gravity, then the resultant force due to the fluid pressure on the surface must be vertically upwards and equal to the weight of the fluid inside the surface. It then follows at once that if a body is wholly or partially immersed in a fluid under gravity, the resultant force due to the fluid pressure on its surface, known as the *upthrust*, must be equal and opposite to the weight of fluid displaced. This result is originally due to Archimedes (ca. 287 – 212 BC).

Clearly, the upthrust acts through the centre of gravity C of the fluid displaced, while the weight of the body acts through the centre of gravity of the body. We wish to find the condition that a body, such as a ship, floats in stable equilibrium. We consider a small rolling displacement, i.e., a rotation about a longitudinal axis of the ship. Then the volume of fluid displaced remains constant, but its shape does not, and so the position of C changes to, say, C'. Clearly, the weight of the ship and the upthrust now form a couple which will tend to right the ship, provided the vertical through C' meets the symmetry plane of the ship at a point M above G. This point is called the *metacentre*. For small

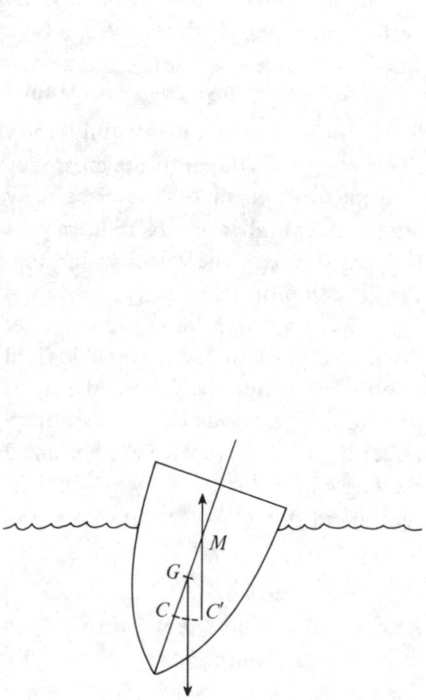

Fig. 7.7

displacements, it can be shown that M is fixed in the ship and is the centre of curvature of the curve CC', but for large displacements, which are important in practice, M may move along the line CG, and this must be taken into account in the design of the hull.

7.4 Fluids in motion

At any given instant of time, the particles of a fluid in motion will have definite velocities. This means that the velocity $\mathbf{v}$ differs from point to point in the fluid, and it is therefore a function of position. We have met vectors that are functions of position before, for instance the gravitational force, and we described this feature by the word *field*. We therefore now refer to the *velocity field* $\mathbf{v}(\mathbf{r})$ of a fluid. An imaginary line drawn in such a field which is tangential to the velocity vector at all points of the line is called a *stream line*. It is also in general a function of time, so that we should write $\mathbf{v}(\mathbf{r}, t)$, but we shall be concerned only with *steady flow*, i.e., flow which is independent of time, and in that case the particles of the fluid move along the stream lines.

The motion of the particles of the fluid must of course be governed by the usual laws of mechanics. We define a *stream tube* as an imaginary tube bounded by stream lines and open at either end. If the cross-sectional area of the tube at a particular point is A, and if the velocity and density of the fluid at that point is $\mathbf{v}$ and ρ, then the mass of fluid crossing A in unit time is $Av\rho$. Now, no fluid can cross the tube walls, and hence if (1) and (2) are two points along the tube, we must have

$$A_1 v_1 \rho_1 = A_2 v_2 \rho_2 = \text{constant.} \tag{7.8}$$

Thus, for an incompressible fluid, the velocity along a stream tube is inversely proportional to the cross-sectional area of the stream tube. This makes it possible to describe the velocity field graphically by letting the number of stream lines per unit cross-sectional area measure the magnitude of the velocity, and the direction of the stream line give the direction of the velocity.

We next consider the relation between the velocity and pressure distributions of an incompressible fluid in motion under gravity. Let a given mass of fluid be displaced a small distance along a stream tube as shown, from position (a) to position (b). [See Fig. 7.11.] The net effect of this is to replace the volume dV of fluid on the left-hand side moving with velocity $\mathbf{v}_1$ by an identical volume dV of fluid on the right-hand side moving with velocity $\mathbf{v}_2$. The change in kinetic energy is, therefore,

$$\Delta T = \tfrac{1}{2}\rho \, dV(v_2^2 - v_1^2). \tag{7.9}$$

At the same time there is a thrust $p_1 dA_1$ moving through distance ds_1 at (1) and a corresponding opposing thrust $p_2 dA_2$ moving through distance ds_2 at (2). Since

$$dA_1 ds_1 = dA_2 ds_2 = dV,$$

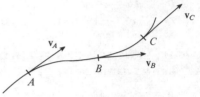

Fig. 7.8

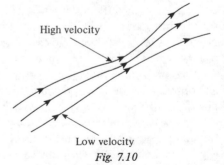

Fig. 7.9

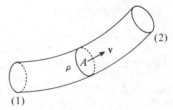

High velocity

Low velocity

Fig. 7.10

the work done on the fluid in the tube is

$$\Delta W = (p_1 - p_2)\mathrm{d}V. \tag{7.10}$$

Finally, if the two points (1) and (2) are at heights z_1 and z_2 above an arbitrary fixed level, then there is a change in potential energy

$$\Delta V = \rho g\,\mathrm{d}V(z_2 - z_1). \tag{7.11}$$

From energy conservation, it follows that

$$\Delta W = \Delta T + \Delta V.$$

$$\therefore\quad p_1 - p_2 = \tfrac{1}{2}\rho(v_2^2 - v_1^2) + \rho g(z_2 - z_1),$$

or, quite generally,

$$p + \rho g z + \tfrac{1}{2}\rho v^2 = \text{constant} \tag{7.12}$$

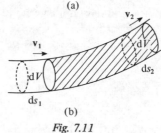

(a)

(b)

Fig. 7.11

for all points along a stream line. This relation between pressure and velocity at any point of a stream line is known as Bernoulli's theorem. [Daniel Bernoulli's (1700–1782) was a member of a family which in three generations produced eight professors of mathematics and astronomy.] It should be noted that p here is the total pressure, and for a liquid in contact with the atmosphere, it includes the atmospheric pressure.

We must distinguish here between *laminar flow*, and *turbulent flow*. In laminar flow, a stream line never crosses its own path. It is therefore possible to fill the whole of the fluid space with overlapping stream tubes, and the constant in Bernoulli's equation must then be the same throughout the fluid. This leads to the apparently paradoxical but correct result that the velocity is greatest where the pressure is least. In turbulent flow, some stream lines form closed loops, known as vortices (singular: vortex). In that case, it is not possible to fill the space with overlapping stream tubes and no general statement regarding the fluid as a whole can be made.

Although Bernoulli's theorem in the form (7.12) is strictly correct only for incompressible fluids, corrections for compressible fluids are small, as long as the velocities in the fluid are not too large. Applications of the effect are numerous. The cross-section of an aeroplane wing is shaped in such a way that the velocity of the air flowing above the wing is higher than that below the wing. (Note the crowding of the stream lines above the wing in Fig. 7.13.) Correspondingly, the pressure is higher below the wing, and the aeroplane experiences a lift. The effect can be increased by tilting the wing, but if this is done beyond a certain limit, eddies form above and behind the wing, which destroy the lift. The aeroplane then stalls, i.e., it drops. Note that the word 'stalls' does not imply that the engine cuts out.

Another consequence, known as the Magnus effect, relates to the flight of a spinning ball. As the air flows past the ball, it clings to some extent to the surface and is carried along with it. Consider a ball moving

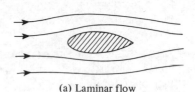

(a) Laminar flow

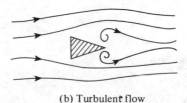

(b) Turbulent flow
Fig. 7.12

Flow past different kinds of obstacles.

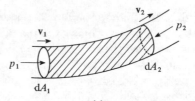

Fig. 7.13

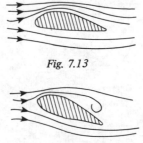

Fig. 7.14

to the left with bottom spin. This is equivalent to a spinning stationary ball in an air stream moving to the right. The effect of the spin on air molecules near the surface of the ball is to speed up those above the ball and slow down those below it. Hence the velocity of the air just above the ball is greater than that just below it. In consequence the ball experiences a lift, just like the aeroplane wing, and it carries further than it would have done without the spin. Top spin has the opposite effect, while spin about a vertical axis leads to sideways drift. The desirable and undesirable results of these effects in tennis, golf and cricket are well known.

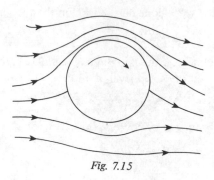

Fig. 7.15

Progress test

1. In steady flow conditions of a fluid,

    A   the fluid is incompressible
    B   the flow is free of vortices
    C   the flow is independent of time
    D   the flow is the same at all points of space.

2. A golf ball is hit in such a manner as to have a clockwise spin about a vertical axis, when viewed from above. As a result the ball will tend to

    A   lift
    B   drop
    C   veer to the left
    D   veer to the right.

7.5   Viscous fluids

The fact that properties of such obviously real things as aeroplane wings and golf balls can be deduced from our consideration of ideal fluids, shows that there are many real fluids that approximate to an ideal fluid. If the shearing forces cannot be neglected, then we say that the fluid has *viscosity*. In that case, laminar flow leads to a tangential force F between successive layers of fluid, which may be taken to be proportional to the area of contact A of the layers and to the rate of change of the velocity of the layers at right angles to the layer surfaces, that is,

$$F = \eta A \frac{\mathrm{d}v}{\mathrm{d}x}, \qquad (7.13)$$

where η is the *coefficient of viscosity*. This formula, originally due to Newton, has been found valid for many fluids. The quantity η has dimensions

$$[\eta] = \mathrm{ML}^{-1}\,\mathrm{T}^{-1}, \qquad (7.14)$$

as can easily be verified, and the common unit of viscosity is still the cgs unit, the poise. In terms of MKS units,

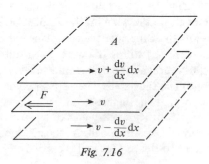

Fig. 7.16

10 poise $= 1$ kg m^{-1} s^{-1}.

The simplest application of (7.13) relates to the laminar flow of a viscous fluid along a tube. As the tube is stationary, so must the outer-most layer of fluid be, while the velocity of the fluid is greatest along the central axis. We consider a length L of the tube of radius R, and investigate the motion of a cylinder of fluid of radius r. The forces on it are due to the pressures at the two ends

$$p_1 \pi r^2 - p_2 \pi r^2$$

and to the viscous drag along the curved surface

$$\eta A \frac{dv}{dr} = \eta 2\pi r L \frac{dv}{dr}.$$

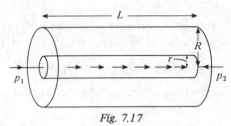

Fig. 7.17

In equilibrium flow, there are no accelerations in the fluid, and hence the resultant force vanishes. Therefore

$$(p_1 - p_2) \pi r^2 + 2\pi r \eta L \frac{dv}{dr} = 0. \tag{7.15}$$

This can be integrated, to yield

$$v = \frac{p_1 - p_2}{4\eta L} (R^2 - r^2). \tag{7.16}$$

Thus, the velocity varies quadratically from the wall to the centre of the tube.

To obtain the rate of flow of fluid through the tube, we note that the part of the fluid that has velocity v is a thin-walled tube of radius r to $r + dr$. The volume passing in time dt through this thin-walled tube is

$$dV = 2\pi r \, dr \, v \, dt.$$

Hence the rate of flow for the whole tube is obtained by integrating dV/dt from $r = 0$ to $r = R$,

$$Q = \int_0^R 2\pi r \, dr \frac{p_1 - p_2}{4\eta L} (R^2 - r^2)$$

$$= \frac{\pi R^4}{8\eta} \frac{p_1 - p_2}{L}. \tag{7.17}$$

This formula, due to Poiseuille, enables us to determine the viscosity of a fluid.

To take the subject further would go beyond the limits of this book, and it must be stressed that many of the really interesting viscous fluids do not even obey Newton's formula (7.13). However, there is one problem which is of considerable interest in atomic physics, namely the force on a small sphere moving through a viscous liquid. This is required in the analysis of Millikan's experiment for the determination of the charge on the electron [see Jackson, chapter 2]. To derive it, we shall use dimensional considerations, which were explained in section 1.9. The force may be expected to depend on (a) the radius R of the sphere, (b) the velocity v of the sphere and (c) the viscosity η of the fluid. Hence, we can write

$$F \propto R^\alpha v^\beta \eta^\gamma,$$

where information on α, β and γ is obtained from the fact that the equation must be dimensionally correct. Hence,

$$\text{MLT}^{-2} = \text{L}^\alpha (\text{LT}^{-1})^\beta (\text{ML}^{-1}\,\text{T}^{-1})^\gamma,$$

and this equation must be separately satisfied for the powers of M, L and T. Hence,

$$1 = \gamma, \qquad 1 = \alpha + \beta - \gamma, \qquad 2 = \beta + \gamma,$$

so that

$$\alpha = 1, \qquad \beta = 1, \qquad \gamma = 1.$$

Hence,

$$F \propto Rv\eta. \tag{7.18}$$

The method of dimensions cannot give us the proportionality constant, which can only be obtained from a complete analysis of the problem. This was done by Stokes (1819–1903), who found that

$$F = 6\pi Rv\eta. \tag{7.19}$$

For a sphere falling under gravity, the velocity increases until the viscous force and buoyancy equal the weight. After that, the velocity remains constant at a value called the *terminal velocity*. If the density of the sphere and of the liquid are ρ and σ, then the terminal velocity v_T is given by

$$\tfrac{4}{3}\pi R^3 \rho g = \tfrac{4}{3}\pi R^3 \sigma g + 6\pi Rv_T \eta.$$

Hence,

$$v_T = \frac{2R^2 g}{9\eta}(\rho - \sigma). \tag{7.20}$$

Progress test

1. The tangential force in viscous laminar flow depends on

   A   the velocity of flow
   B   the velocity gradient along the flow direction
   C   the velocity gradient at right angles to the flow direction
   D   the acceleration of the fluid.

7.6 Problems

7.1. A container of water is moved along a frictionless horizontal table with an acceleration f. (a) Describe the free surface. (b) If the table is now tilted and the container is allowed to slide down freely, describe the free surface.

7.2. It is shown in chapter 8 that for adiabatic conditions, i.e., when no heat enters or leaves a system, the relation between the pressure and density of a gas is

$$p\rho^{-\gamma} = \text{constant},$$

where $\gamma = 1\cdot4$ for air. Show that, for an adiabatic atmosphere,

$$p = p_0\left(1 - \frac{\gamma-1}{\gamma}\frac{\rho_0 gz}{p_0}\right)^{\frac{\gamma}{\gamma-1}},$$

and that the pressure vanishes at a height of 28 km. [Normal air pressure and density at ground level are

$$p_0 = 1\cdot03 \times 10^5 \text{ Nm}^{-2}, \qquad \rho_0 = 1\cdot29 \text{ kg m}^{-3}.]$$

7.3. Show that, for an isothermal atmosphere, the pressure at a height of 28 km is three per cent of that at ground level.

7.4. A wooden hemisphere floats on water with its flat side uppermost. Show that the equilibrium is stable.

7.5. The wing loading on a modern commercial aircraft is of the order of 400 kgf m^{-2}. Show that, if the speed of the aircraft is 800 km h^{-1}, the difference in the velocity of airflow above and below the wing is of the order of five per cent of the velocity of the aircraft.

*7.6. An open vessel filled with water has a tap at a depth h below the surface of the water. Show that, when the tap is opened, the speed with which the water emerges is equal to the speed of a body falling from rest through a height h. [Hint: Use Bernoulli's theorem and note that when the tap is open, the water in the tap is in contact with the atmosphere.]

7.7. A bubble of air, 1 mm in diameter, rises in water. What is the terminal velocity? The viscosity of water at room temperature is 0·010 poise.

8. The molecular theory of matter

8.1 Introduction

In the last two chapters we treated certain properties of matter in bulk. We did this from what is called a *macroscopic view*, by which is meant that we were not concerned about the way in which the bulk matter was made up from molecules. In fact, we treated it as a continuous homogeneous substance with certain bulk properties, such as rigidity, elasticity, pressure or viscosity. In this chapter we wish to take a *microscopic view*, i.e., we shall try to understand how some of the bulk properties arise from the behaviour of the molecules that make up bulk matter. At first, it will be adequate to use a model in which the molecules are considered to be point particles that interact with each other through *intermolecular forces*, which we do not need to specify in any detail. Later, we shall be forced to look at the structure of the molecules themselves, in order to account for some of the bulk properties which we are investigating.

8.2 Intermolecular forces

The most obvious bulk property of matter is that matter can exist—broadly speaking—in three different states, solid, liquid and gaseous. Related to this is the fact that, with rare exception, solids expand slightly on liquefying, and, without exception, liquids expand considerably when turning into gases. Let us start by considering gases, which lack rigidity and cohesion and are easily compressible. These properties are readily accounted for, if we think of a gas as an assembly of molecules which is such that the spaces between the molecules are substantially larger than the spaces occupied by the molecules. The molecules are then so far apart that the forces between them are very small, and the molecules are able easily to alter their relative position within the volume of the gas. In the diagram, the molecules are shown

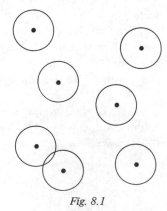

Fig. 8.1

as points, and the circles indicate the distance over which the intermolecular forces are effective. This is called the *range* of the force, and beyond it the effect of the force may be assumed to be negligible.

A liquid still lacks rigidity, but it has cohesion and it is almost incompressible. This must mean that in a liquid the molecules are so closely packed that it is almost impossible for them to approach each other more closely, but at the same time it is still possible for the molecules to move freely relative to each other. At this distance between the molecules, the intermolecular force must be attractive, to account for the cohesion of a liquid. Next, in a solid, the density is hardly greater than in a liquid, so that the molecules are hardly more closely packed than in a liquid. On the other hand, a solid has rigidity, which means that the molecules in a solid essentially maintain their positions relative to each other. The very small change in the average separation of the molecules in going from liquid to solid must therefore lead to a considerable increase in the intermolecular force. From Figs. 8.2 and 8.3, it is apparent that in our model the increase in density between liquid and solid is actually in part due to the more ordered arrangement of molecules in the solid, which reduces the spaces between the molecules. Such an orderly arrangement is reflected on the macroscopic scale in the crystalline nature of many solids [R. L. Fullman, 'The growth of crystals', *SA* 260, March 1955, p. 74]. Finally, solids are effectively incompressible, so that the application of even very great forces to solids in bulk does not bring the molecules significantly closer to each other. This means that at some distance, called the equilibrium distance, the intermolecular force must change rapidly from an attraction to a repulsion.

The above qualitative considerations enable us to sketch a graph of the intermolecular force F as a function of distance r between molecules. It vanishes at large distances, but becomes rapidly attractive, i.e., negative, at a distance which we called the range. After passing through a (negative) maximum it even more rapidly becomes zero at the equilibrium distance and then large and repulsive, i.e., positive, at small distances.

We can even obtain reasonable estimates of the equilibrium distance and of the range from our knowledge of the mass of a hydrogen atom and of typical densities of solids and gases. The mass of a hydrogen atom is about $1 \cdot 67 \times 10^{-27}$ kg, and that of a water molecule is 18 times as large. Hence, 1 kg of water contains $3 \cdot 3 \times 10^{25}$ molecules of water. As the density of water is 1000 kg m^{-3}, this means that each molecule occupies a space of $3 \cdot 0 \times 10^{-29}$ m^3. The intermolecular distance must be of the order of the cube root of this volume, i.e., about 3×10^{-10} m. This, then, is the equilibrium distance. A maximum estimate of the range is obtained from the fact that the density of water vapour is 1600 times less than that of liquid water. The distances between molecules are therefore greater by a factor

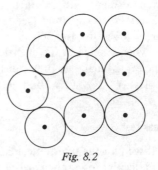

Fig. 8.2

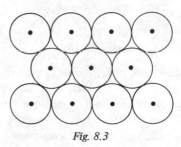

Fig. 8.3

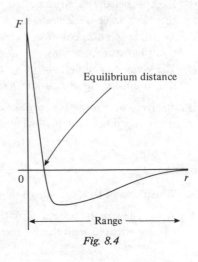

Fig. 8.4

$1600^{1/3} \simeq 12$. Thus, the range of the intermolecular force is of the order of 4×10^{-9} m. These very crude estimates are in fact in substantial agreement with accurate determinations, such as those obtained from X-ray crystallography [see Jackson, chapter 8], and the shape of the force curve, too, is much as would be derived from fundamental theory when applied to the electric interactions between the electrons and nuclei that make up molecules [B. V. Derjaguin, 'The force between molecules', *SA* 266, July 1960, p. 47].

Progress test

1. If we plot the intermolecular potential instead of the intermolecular force, then at the equilibrium point, the potential is

   A   a maximum
   B   a minimum
   C   zero
   D   none of the above.

8.3   Ideal gases

Just as we found it too difficult to deal with the special macroscopic properties of real fluids, so it will not be possible to apply our knowledge of the intermolecular force to real substances on the microscopic scale. We shall therefore consider an *ideal gas* as an assembly of molecules with the following properties:

(a)   The molecules move freely throughout the volume of the gas, except when they make collisions with each other or with the walls of the containing vessel.
(b)   Such collisions are to be treated as perfectly elastic.
(c)   Except that they lead to collisions, the intermolecular forces are therefore to be neglected.
(d)   The time spent in collisions is negligible compared with the time spent between collisions.
(e)   The volume occupied by the molecules is negligible compared with the volume of the containing vessel.

It is intuitively obvious that real gases at very low densities approximate to ideal gases, and this is indeed what we shall find.

Progress test

1. Which one of the following is not a property of an ideal gas?

   A   The volume occupied by the molecules is negligible compared with the volume of the vessel which contains the gas

B   At any given temperature, the molecules have a range of velocities

C   The number of collisions per second is large

D   The time spent in collisions is negligible compared with the time spent between collisions.

Height (m)	Number
1·30–1·31	2
1·31–1·32	0
1·32–1·33	4
1·33–1·34	7
1·34–1·35	3
1·35–1·36	2
1·36–1·37	2

8.4   Excursion into statistics

Our task now is to deduce certain macroscopic properties on the basis of the microscopic behaviour of the individual molecules. Clearly, it would be impossible to do this by considering the motion of each molecule separately, neither is this desirable, since the macroscopic properties must be due to the *average* behaviour of the very large number of molecules that make up even the smallest quantity of matter that we ever deal with macroscopically in the laboratory. To obtain these average properties, we have to use the methods of statistics.

If we measured the heights of, say, 20 children of a certain age to the nearest 0·01 m, we might obtain the result shown in the table. We could plot this as a histogram which would show the frequency F with which a given height occurred, and this would be known as a *frequency distribution*. If we then divide the numbers for each height by the total number of children, we obtain another histogram. The height of each column is now a measure of the *probability* that a child in the group, selected at random, has a height between the limits indicated by the width of that column.

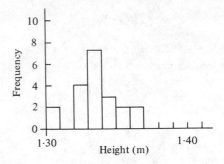

The length of the intervals, in this case 0·01 m, into which we have divided the total range over which our measurements extend, is rather arbitrary. We therefore introduce the concept of a *probability density*, which is the probability per interval of unit length, where the unit in our illustration is 1 m. This concept is particularly useful when the length of interval becomes much smaller than the unit length. Let us consider quite generally a population of N members, in which we measure a property x. Then, if F_i members have a measure between x_i and $x_i + dx_i$, and if the probability density at x_i is P_i, then the probability of finding a measure between x_i and $x_i + dx_i$ is

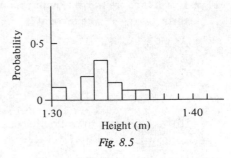

Fig. 8.5

$$P_i \, dx_i = F_i/N. \tag{8.1}$$

In this way, the probability density P is given by the y-co-ordinate on a graph, but the probability itself by an area under the graph. It follows from (8.1) that

$$\sum_i P_i \, dx_i = 1, \tag{8.2}$$

since obviously, $\Sigma F_i = N$.

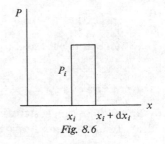

Fig. 8.6

To obtain the average value $\bar{x}$ of all the N measurements of the property x, we note again that there are F_i measurements that yield a value between x_i and $x_i + dx_i$. Hence,

$$N\bar{x} = \sum_i x_i F_i$$

or, using (8.1),

$$\bar{x} = \sum_i x_i P_i \, dx_i. \tag{8.3}$$

In the limit, as $dx \to 0$, we obtain a continuous *probability distribution* $P(x)$ which is such that the probability of a measurement giving a result between x and $x + dx$ is $P(x)\,dx$, and the total probability of obtaining any measurement whatsoever is

$$\int P(x)\,dx = 1. \tag{8.4}$$

Thus unity is a measure of *certainty*. For the mean, we obtain from (8.3)

$$\bar{x} = \int x P(x)\,dx, \tag{8.5}$$

Similarly if we wanted, say, the average of the measurement of the property x^2, which we shall need below, then

$$\overline{x^2} = \int x^2 P(x)\,dx. \tag{8.6}$$

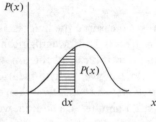

Fig. 8.7

Solution

1. (C) How large is 'large'? There must be a comparison, if the statement is to mean anything. Even then, it is not relevant.

Solution

1. (B) The force is the negative gradient of the potential.

Progress test

1. If two dice are thrown simultaneously, then the probability of a throw giving a score of eight is

    A $\frac{5}{36}$
    B $\frac{1}{6}$
    C $\frac{7}{36}$
    D $\frac{2}{3}$.

2. If two dice are thrown simultaneously, then the average score for a throw is

    A   6
    B   7
    C   8
    D   none of the above.

8.5 The pressure of a gas

We now apply those simple statistical ideas to an assembly of molecules in an ideal gas, in order to calculate the pressure of the gas. We consider

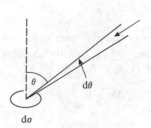

Fig. 8.8

a molecule of mass m incident at an angle between θ and $\theta + d\theta$ to the normal on an element $d\sigma$ of the wall of the vessel. We make the reasonable assumption that molecules move with equal probability in all directions, both towards and away from $d\sigma$. Then, the probability $P(\theta)$ of molecules approaching $d\sigma$ at an angle between θ and $\theta + d\theta$ is equal to the ratio of that area of a unit sphere centred on $d\sigma$ which these molecules must cross, to the total area of the unit sphere. Hence

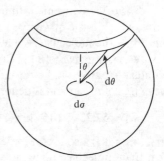

$$P(\theta) = \frac{2\pi \sin \theta \, d\theta}{4\pi} = \tfrac{1}{2}\sin \theta \, d\theta. \tag{8.7}$$

Fig. 8.9

We next restrict ourselves to those molecules that have speeds between C and $C + dC$, and let the probability for this be $P(C)\,dC$. We do not know, at present, what this probability is, but shall evaluate it in section 8.8. Finally, the number of molecules that hit $d\sigma$ in unit time is the number contained in a skew cylinder of length C on $d\sigma$ at angle θ. [Fig. 8.10.] This is $NC\cos\theta\,d\sigma$, where N is the number of molecules in unit volume of the gas.

Bringing together all the above considerations, we find that the number of molecules hitting $d\sigma$ in unit time at an angle between θ and $\theta + d\theta$ with speeds from C to $C + dC$ is

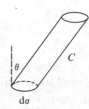

$$NC\cos\theta\,d\sigma \cdot \tfrac{1}{2}\sin\theta\,d\theta\,P(C)\,dC. \tag{8.8}$$

Fig. 8.10

Now each of these molecules is reflected elastically by $d\sigma$, and therefore the total change of the component of momentum perpendicular to $d\sigma$ for each molecule is [Fig. 8.11]

$$2mC\cos\theta. \tag{8.9}$$

The force on $d\sigma$ is the total change of momentum in unit time and is therefore obtained by multiplying together expressions (8.8) and (8.9), while the pressure is obtained by dividing this quantity by $d\sigma$. Finally, to obtain the total pressure, we must integrate over all possible angles θ and all possible speeds C. This gives

Fig. 8.11

$$p = \int_0^\infty \int_0^{\pi/2} 2mC\cos\theta \; NC\cos\theta \; \tfrac{1}{2}\sin\theta\,d\theta\,P(C)\,dC. \tag{8.10}$$

We at once perform the θ-integration and obtain

$$p = \tfrac{1}{3}mN \int_0^\infty C^2 P(C)\,dC. \tag{8.11}$$

If we compare this with (8.6), we find that the integral is the average value of C^2, so that

$$p = \tfrac{1}{3}mN\overline{C^2}. \tag{8.12}$$

The quantity $\overline{C^2}$ is called the *mean square velocity*, and $\sqrt{\overline{C^2}}$ is the *root mean square velocity*.

It is often convenient to consider one *kilomole* (kmol) of a substance, which is the quantity of the substance that has a mass of M kg, where M is the molecular weight of the substance. Thus, one kilomole of water has a mass of 18 kg. Clearly one kilomole of any substance contains the same number of molecules. This number is called Avogadro's number and is denoted by N_A. Its value is

$$N_A = 6 \cdot 023 \times 10^{26} \text{ kmol}^{-1}. \tag{8.13}$$

Sometimes it is also useful to define the kilomole in terms of entities other than molecules. Thus, an atomic kilomole of a substance has a mass of A kg, where A is the atomic weight of the substance. Since the atomic weight of hydrogen is $1 \cdot 008$, it follows that the mass of 1 atom of hydrogen is

$$\frac{1 \cdot 008}{6 \cdot 023 \times 10^{26}} = 1 \cdot 67 \times 10^{-27} \text{ kg}.$$

This was the number used in problem 3.10. The very large value of N_A was also referred to in section 6.1.

If one kilomole of a particular gas occupies a volume V, then it follows from (8.12) that

$$pV = \tfrac{1}{3} m N_A \overline{C^2}. \tag{8.14}$$

Now the kinetic energy of a molecule moving with velocity C is

$$E_K = \tfrac{1}{2} m C^2. \tag{8.15}$$

Hence the average kinetic energy of the molecules is

$$\bar{E}_K = \tfrac{1}{2} m \overline{C^2}, \tag{8.16}$$

and the total kinetic energy of all the molecules in a kilomole of gas is

$$U = \tfrac{1}{2} m N_A \overline{C^2}. \tag{8.17}$$

Substituting this quantity in (8.14), we have

$$pV = \tfrac{2}{3} U. \tag{8.18}$$

This equation links the pressure and volume of the gas, which are macroscopic properties with the total kinetic energy of the molecules, a microscopic property.

Progress test

1. Doubling which one of the following does not double the pressure of an ideal gas?

    A   The mass of each of the gas particles
    B   The average velocity of the gas particles

Solutions

1. (A) There are 36 possible results, when throwing two dice and five of these yield a score of 8.

2. (B) Work out in how many ways one can score 2, 3, etc. and then average.

C   The density of the gas
D   The number of gas particles in a given volume.

2. The mean square velocity of the velocities of a number of
particles is

A   the average of the squares of the velocities
B   the square of the average of the velocities
C   the square of the most probable velocity
D   none of these.

3. Avogadro's number is the number of molecules in

A   1 kg
B   1 kilomole
C   the equivalent weight
D   none of these.

8.6   The concept of temperature

In order to compare our theory with experiment, we must introduce
the concept of *temperature*. This is a new concept, which cannot be
defined through anything contained in mechanics. We therefore
proceed in the same manner as when we introduced the concept of
force in chapter 1. We again quote Sommerfeld, this time in his
Thermodynamics, Academic Press, 1964:

'We have a qualitative measure for temperature in our sense of
touch, and a quantitative one, however arbitrary, in every thermo-
meter.'

Temperature is a property of the state of a system with the feature that
if two systems are in contact with each other for a sufficiently long
time, then their temperatures become the same.

8.7   Temperature and kinetic energy

We next show that the total kinetic energy U of a kilomole of gas, which
we defined in (8.17), can be linked to our new concept of temperature.
We consider a vessel having two different kinds of molecules, N_1 per
unit volume of mass m_1, and N_2 per unit volume of mass m_2. If one of
each of these collide, with velocities C_1 and C_2 respectively, then the
velocity of (1) relative to (2) is

$$\mathbf{v} = \mathbf{C}_1 - \mathbf{C}_2. \tag{8.19}$$

As a result of an elastic collison, this relative velocity is unchanged in
magnitude, but turned through some angle. This is most easily seen by
considering the collision in the CM-frame, as was done in Fig. 4.5.

Hence, the relative velocities of the molecules are constantly changed by the repeated collisions of the molecules, and it is reasonable to assume that on average all directions of relative velocities eventually become equally likely, whatever may have been the initial conditions. Now the velocity of the centre of mass of the two colliding molecules is

$$V = \frac{m_1 C_1 + m_2 C_2}{m_1 + m_2}, \tag{8.20}$$

and what we are saying is that, because of the randomizing effect of the collisions, all directions of v relative to V are equally likely. This means, that on average

$$\langle v \cdot V \rangle_{AV} = 0. \tag{8.21}$$

Substituting from (8.19) and (8.20), we obtain

$$\left\langle \frac{m_1 C_1^2 - m_2 C_2^2 + (m_2 - m_1) C_1 \cdot C_2}{m_1 + m_2} \right\rangle_{AV} = 0. \tag{8.22}$$

But C_1 and C_2 also are velocities with quite arbitrary directions, so that

$$\langle C_1 \cdot C_2 \rangle_{AV} = 0. \tag{8.23}$$

Hence, we have

$$\langle m_1 C_1^2 \rangle_{AV} = \langle m_2 C_2^2 \rangle_{AV}, \tag{8.24}$$

or, in our previous notation

$$\tfrac{1}{2} m_1 \overline{C_1^2} = \tfrac{1}{2} m_2 \overline{C_2^2}. \tag{8.25}$$

We thus have arrived at the remarkable result that if two different gases are in equilibrium with each other, then the average kinetic energy is the same for both kinds of molecules.

The same result holds true, if the two gases are separated, by a movable piston in a cylindrical vessel. For if this were not so, then it would follow from our expression (8.12) for the pressure of a gas that the pressures on the two sides of the piston would be different and the piston would move. In this case, the system was not initially in equilibrium. It may be noticed that we have assumed in this proof that the number N of molecules per unit volume is the same for both gases. This assumption, known as Avogadro's hypothesis, has been amply verified in practice.

Returning now to our result (8.17) for the total kinetic energy, we see that if a kilomole of each of two gases are in equilibrium, then their total kinetic energies are the same, and conversely. But from our discussion of temperature, it follows that two gases in equilibrium with each other are at the same temperature. Hence the total kinetic energy is a property of the temperature and independent of the particular gas. Conversely, temperature is a function of the total kinetic energy of the

Solutions

1. (B) Doubling the velocity quadruples the pressure.

2. (A) See (8.6).

3. (B) This is bookwork.

Fig. 8.12

system, and this, in principle, gives us a means of obtaining a measurement of temperature by measuring the kinetic energy of the system. We therefore now *define* temperature quantitatively as being proportional to the total kinetic energy of an ideal gas, that is,

$$U \propto T. \tag{8.26}$$

In terms of the usual absolute temperature scale, we define the proportionality constant through

$$U = \tfrac{1}{2}mN_A\overline{C^2} = \tfrac{3}{2} RT, \tag{8.27}$$

where T is the temperature in K and R is called the *universal gas constant*, which has the value

$$R = 8317 \text{ J kmol}^{-1}\text{K}^{-1}. \tag{8.28}$$

It is often convenient to use another constant k, called *Boltzmann's constant* defined through

$$k = R/N_A = 1 \cdot 38 \times 10^{-23} \text{ J K}^{-1}. \tag{8.29}$$

In terms of this constant, the average kinetic energy of a single molecule is

$$\tfrac{1}{2}m\overline{C^2} = \tfrac{3}{2} kT. \tag{8.30}$$

Finally, we substitute (8.27) into the expression (8.18), which connects the pressure and volume of a kilomole of an ideal gas, and obtain

$$pV = RT. \tag{8.31}$$

This is the well-known equation of state for an ideal gas and incorporates both Boyle's and Charles's laws. We have therefore arrived at the gas laws by an application of Newtonian mechanics to the motion of the gas molecules. For this reason the theory is called the *kinetic theory of gases*. It must, however, be stressed that we have not *deduced* the gas laws from Newtonian mechanics, since in the process we have introduced ideas of probability, which are not contained in Newtonian mechanics.

Progress test

1.　In a mixture of two ideal gases, of particle masses m_1 and m_2 with $m_1 > m_2$, the average kinetic energies $\overline{T}_1$ and $\overline{T}_2$ are always such that

　　A　　$\overline{T}_1 > \overline{T}_2$

　　B　　$\overline{T}_1 = \overline{T}_2$

　　C　　$\overline{T}_1 < \overline{T}_2$

　　D　　none of these.

2.   For a real gas, Boyle's law is valid in the limit of

    A   high pressure
    B   low pressure
    C   low temperature
    D   small volumes.

8.8   The distribution of velocities in a gas

We now return to the problem of the distribution of velocities in a gas, which we raised in section 8.5. This problem was first tackled by Maxwell (1831–1879) and we shall follow his proof.

Let the velocity $\mathbf{C}$ of a molecule have components u, v, w along an arbitrary set of co-ordinate axes, and let the probability that the first component has a value between u and $u + du$ be

$$g(u)\,du. \tag{8.32}$$

The same function will apply to the other components, since all directions are equivalent. The probability of finding the magnitude of the velocity between C and $C + dC$ is then

$$g(u)\,g(v)\,g(w)\,du\,dv\,dw. \tag{8.33}$$

however, this cannot depend separately on u, v and w, but only on C. Hence it must be given by a function of C only. We can write $du\,dv\,dw$ as a volume element in velocity space, and in polar co-ordinates this becomes

$$du\,dv\,dw = 4\pi C^2\,dC$$

for the volume element of velocities between C and $C + dC$ in *any* direction. Hence, we can write the probability of finding the velocity between C and $C + dC$ as

$$P(C)\,dC, \tag{8.34}$$

where

$$\frac{P(C)}{4\pi C^2} = g(u)\,g(v)\,g(w) \tag{8.35}$$

and

$$C^2 = u^2 + v^2 + w^2. \tag{8.36}$$

We now differentiate with respect to u:

$$\frac{d}{dC}\left[\frac{P(C)}{4\pi C^2}\right]\frac{dC}{du} = \frac{dg(u)}{du}\,g(v)\,g(w), \tag{8.37}$$

where

$$\frac{dC}{du} = \frac{u}{C}.$$

Dividing (8.37) by (8.35), we have

$$\frac{4\pi C}{P(C)} \frac{d}{dc}\left[\frac{P(C)}{4\pi C^2}\right] = \frac{1}{ug(u)}\frac{dg}{du}, \qquad (8.38)$$

where the right-hand side depends only on u, while the left-hand side depends on u, v, w through C. This is only possible if both sides are equal to a constant, which we denote by -2β.

$$\therefore \quad \frac{1}{ug(u)}\frac{dg}{du} = -2\beta. \qquad (8.39)$$

Integrating, we have

$$\ln g(u) = \alpha - \beta u^2,$$

or

$$g(u) = a\,e^{-\beta u^2} \qquad (8.40)$$

where $a = e^{-\alpha}$ is a constant. It is determined from the fact that the total probability must be unity, that is,

$$\int_{-\infty}^{\infty} g(u)\,du = 1. \qquad (8.41)$$

Using the so-called Gaussian integral

$$\int_{-\infty}^{\infty} e^{-x^2}\,dx = \sqrt{\pi}, \qquad (8.42)$$

we find that

$$g(u) = \sqrt{\frac{\beta}{\pi}}\,e^{-\beta u^2}. \qquad (8.43)$$

Thus, $g(u)$ is a distribution that gives maximum probability for $u = 0$, as might have been expected, since positive and negative values of u are uqually likely. Clearly, β must be positive, and in that case very large values of u are improbable.

To determine β, we evaluate the mean square value,

$$\overline{u^2} = \int_{-\infty}^{\infty} u^2 g(u)\,du = \frac{1}{2\beta}, \qquad (8.44)$$

with the help of (8.42). Hence,

$$\overline{C^2} = \overline{u^2} + \overline{v^2} + \overline{w^2} = \frac{3}{2\beta}. \qquad (8.45)$$

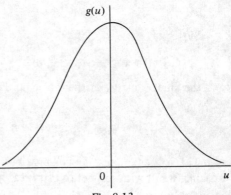

$g(u)$

0 u

Fig. 8.13

But we know from (8.30) that

$$\overline{C^2} = 3kT/m. \tag{8.46}$$

Hence

$$\beta = m/(2kT). \tag{8.47}$$

This can now be substituted in (8.35), and gives

$$P(C) = 4\pi C^2 \left(\frac{m}{2\pi kT}\right)^{3/2} \exp\left[-\frac{m}{2kT}(u^2 + v^2 + w^2)\right]. \tag{8.48}$$

But

$$\tfrac{1}{2}m(u^2 + v^2 + w^2) = \tfrac{1}{2}mC^2 = E_K, \tag{8.49}$$

the kinetic energy of the molecule. This yields

$$P(C) = 4\pi C^2 \left(\frac{m}{2\pi kT}\right)^{3/2} e^{-E_K/(kT)}. \tag{8.50}$$

This is the famous Maxwell distribution law of velocities. By differentiating (8.50) with respect to C it is easily seen that the probability has a maximum when

$$E_K = kT. \tag{8.51}$$

The law has been verified by Estermann, Simpson and Stern (1947), who used an atomic beam technique, in which gas molecules from a heated gas in an oven escaped through a hole in the oven wall, and their speeds were then measured by observing their trajectories under gravity. Although the distance dropped is only a fraction of 1 mm, this could be measured accurately, since the diameter of the detecting wire was only 0·02 mm.

The distribution law, which we have derived, shows that at a given temperature the velocities of the molecules may differ widely, but that the most probable value corresponds to the kinetic energy $E_K = kT$. Because of the skewness of the distribution, this is not, however, the average kinetic energy of the molecules. This, as we know from (8.30), is $\tfrac{3}{2}kT$.

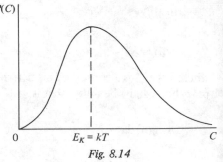

Fig. 8.14

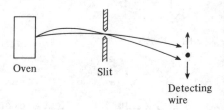

Oven   Slit

Detecting wire

Fig. 8.15

Progress test

1. For a gas molecule, the quantity $E = kT$ gives the

    A   maximum kinetic energy
    B   minimum kinetic energy
    C   average kinetic energy
    D   most probable kinetic energy.

2.   The ratio of the most probable velocity to the root mean square
      velocity for the molecules of a gas at a given temperature is

      A $\frac{2}{3}$

      B $\sqrt{\frac{2}{3}}$

      C $\sqrt{\frac{3}{2}}$

      D $\frac{3}{2}$.

8.9   Polyatomic gases

So far, we have been dealing with monoatomic gases, which could be
treated as mass points, so that the total energy of the gas was given by
the translational kinetic energy of the molecules. We now know that this
can be increased by increasing the temperature of the gas, i.e., by
supplying energy in the form of heat. For polyatomic gases, we must
consider the molecules as extended structures, which can rotate, and
parts of which may vibrate relative to each other. We then have
additional forms of energy, namely rotational kinetic energy, vibrational
kinetic energy and vibrational potential energy, since the possibility of
vibrations implies the existence of forces inside the molecule. The
question arises as to how energy that is supplied to the gas, say, in the
form of heat energy, distributes itself over these various forms of energy.

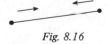

Fig. 8.16

     The answer to this question lies in a famous theorem, which we
can only state, the theorem of the *equipartition of energy*. It is due to
Boltzmann (1844–1906) and states that, in a system that can be treated
statistically, each degree of freedom of the system contributes a
quantity $\frac{1}{2}kT$ to the total energy. Thus, the total energy of a kilomole
of molecules, each of which has n degrees of freedom is

$$U = \tfrac{1}{2}nRT. \tag{8.52}$$

The idea of degrees of freedom was discussed in section 6.3, where we
found that a point particle had three and a rigid body six degrees of
freedom. The extra three degrees correspond to rotations about three
mutually perpendicular axes. Thus, we would expect $n = 3$ for monatomic
molecules, and $n = 6$ for polyatomic molecules, provided these can be
treated as rigid bodies. Between these there are diatomic molecules,
consisting of two point masses a fixed distance apart. These have only
two rotational degrees of freedom, so that $n = 5$, since for two point
masses there can be no rotation about the line joining the two masses.
Both diatomic and polyatomic molecules may have internal degrees of
freedom, corresponding to deformations. For instance, a diatomic
molecule may vibrate along the line joining the two atoms. Such a
vibration actually corresponds to two degrees of freedom from the
point of view of the equipartition theorem, since it leads not only to a

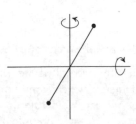

Fig. 8.17

vibrational kinetic energy, but also to an equal vibrational potential energy.

While the total energy of a gas at a given temperature therefore increases with the number of degrees of freedom of the gas molecules, the effect on pressure is still only due to that part of the kinetic energy which is translational. Referring back now therefore to (8.18), which related the pressure p and volume V of a kilomole of gas to the total energy U, we must now generalize this equation to

$$pV = \frac{2}{n} U, \qquad (8.53)$$

since if $n > 3$, the translational kinetic energy is $3/n$ times the total energy. It is conventional to define a quantity γ through

$$\gamma = 1 + 2/n. \qquad (8.54)$$

We then have

$$pV = (\gamma - 1) U = RT \qquad (8.55)$$

for the equation of state of a gas.

Progress test

1. The mean translational kinetic energy of a molecule is

    A $\frac{1}{2}kT$

    B kT

    C $\frac{3}{2}kT$

    D   none of these.

2. If the total energy of a gas is U, and the molecules of the gas have 5 degrees of freedom, then pV equals

    A $\frac{1}{5} U$

    B $\frac{2}{5} U$

    C $\frac{3}{5} U$

    D $5 U$

8.10 The specific heats of gases and solids

Now that we have linked the pressure and volume of a mass of a gas to its total energy, we can apply our theory to the problem of the heat capacity of a gas. We define the *specific heat* of a substance as the energy required to heat one kilomole of the substance through $1°K$. This definition is adequate for solids and liquids, but for gases, which expand significantly when heated, we have to be more precise. We therefore define two specific heats C_p and C_V, depending on whether

Solutions

1. (D)   It is the energy at which the Maxwell distribution gives a maximum probability for the velocity.

2. (B)   The ratio of the corresponding kinetic energies is $\frac{3}{2}$.

we keep the pressure or the volume constant during the heating process.

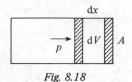

Fig. 8.18

When a gas expands at constant pressure, it does work on its surroundings and hence loses energy. Consider the gas contained in a cylinder with a freely moving piston of cross-sectional area A. The work done by the gas in moving the piston through distance dx is

$$dW = pA\,dx \tag{8.56}$$

and the increase in volume is

$$dV = A\,dx. \tag{8.57}$$

Hence the work done by the gas is

$$dW = p\,dV. \tag{8.58}$$

Since the expansion takes place at constant pressure, we obtain from (8.55) that, for a kilomole of gas,

$$p\,dV = R\,dT. \tag{8.59}$$

Now, it follows from the definition of specific heat that the energy required to heat one kilomole of gas through temperature difference dT is $C_p dT$ when the pressure is kept constant, and $C_V\,dT$ when the volume is kept constant. The difference between these is clearly due to the work done in the first case, when the gas expands at constant pressure. Hence, from (8.59),

$$C_p\,dT - C_V\,dT = R\,dT,$$

or

$$C_p - C_V = R. \tag{8.60}$$

Further, in the second case, no work is done by the gas and so $C_V\,dT$ must equal the increase in the total energy of the gas, dU. This is again obtained from (8.55), that is,

$$(\gamma - 1)\,dU = R\,dT. \tag{8.61}$$

Thus,

$$C_V = \frac{R}{\gamma - 1}. \tag{8.62}$$

Combining (8.60) and (8.62), we find that

$$C_p/C_V = \gamma, \tag{8.63}$$

so that the ratio of the two specific heats of a gas yields the number of degrees of freedom of the molecules of the gas, a truly remarkable result.

The ratio of the specific heats has been measured for many gases and at many temperatures, and typical results are shown in the table. We deal in turn with monatomic, diatomic and polyatomic molecules.

(a) For monatomic molecules we find $n = 3$ at all temperatures. It thus appears that these can indeed be treated as point particles.

(b) For diatomic molecules we find $n = 5$ only at intermediate temperatures. At higher temperatures, n approaches 7, which may indicate the appearance of vibrational kinetic and potential energies, which are absent at lower energies. Even more surprising, the number of degrees of freedom for the hydrogen molecule decreases to $n = 3$ near the point of liquefaction.

(c) We list two examples of polyatomic molecules, which show that a simple molecule, such as NH_3, may indeed be treated as approximately a rigid body, but that more complicated molecules, such as C_2H_6, have internal degrees of freedom.

Gas	$T(°K)$	γ	n
He	93	1·66	3
Kr	292	1·68	3
Ar	288	1·67	3
H_2	273	1·40	5
O_2	373	1·40	5
Cl_2	288	1·36	5·5
Br_2	573	1·32	6·3
I_2	458	1·30	6·7
H_2	80	1·66	3
NH_3	288	1·31	6·5
C_2H_6	288	1·22	9

It would appear that only at temperatures far removed from the point of liquefaction are all possible degrees of freedom operative, and that as the temperature is lowered, more and more degrees of freedom get 'frozen out'. This is in total disagreement with the statement of the equipartition of energy, according to which the total energy available is equally distributed over *all* possible degrees of freedom, irrespective of temperature.

A similar situation arises for solids. For atoms bound in a crystal lattice, we have three degrees of freedom for each atom. But as the atoms vibrate about their mean positions under the forces due to other atoms, there is an equal potential energy contribution. Thus, effectively, $n = 6$, and the total energy for each atom is $3kT$. Hence, from (8.62), the specific heat for all solids is

$$C_V = 3R.$$

This law, discovered empirically by Dulong and Petit, is found to be generally valid at room temperatures, but just as in the case of gases, C_V decreases with temperature, so that degrees of freedom appear to be 'frozen out'.

The difficulties regarding the specific heats already worried Maxwell in 1859. Kelvin in 1884 called them 'a cloud over the dynamical theory of heat', and in 1890 Jeans was puzzled by what he in fact had called the 'freezing out' of certain kinds of motion. To imagine that the physicists of the 19th century were complacent to the point of blindness regarding their theories, is quite wrong.

The answer came through the quantum theory of Planck (1859–1947), according to which energy is not infinitely divisible. The equipartition theorem cannot then be valid down to the absolute zero of temperature, and is only strictly valid in the limit of high temperatures. The quantum theory in fact completely accounts for the anomalies

Solutions

1. (C) Each degree of freedom gives $\frac{1}{2}kT$. For translational motion there are three degrees of freedom.

2. (C) $pV = \frac{2}{3}kT$ and $U = \frac{3}{2}kT$ in this case.

mentioned above, but that is a story not for this book [K. K. Darrow, 'The quantum theory', *SA* 205, March 1952, p. 47].

Progress test

1. For the specific heats C_p and C_V of a gas we always have

    A $C_p < C_V$
    B $C_p = C_V$
    C $C_p > C_V$
    D   none of these.

2. The specific heat of a solid at very low temperatures is always

    A   greater than that at room temperature
    B   equal to that at room temperature
    C   less than that at room temperature
    D   none of these.

8.11 Problems

8.1. Show that an intermolecular potential of the form

$$V(r) = -\frac{A}{r^6} + \frac{B}{r^{12}}$$

satisfies the qualitative features established in section 8.2.

8.2. Determine the root mean square velocities of
    (a)   hydrogen and oxygen molecules at $300°K$,
    (b)   oxygen molecules at $-100°C$, $20°C$, $1000°C$.

8.3. Calculate the following:
    (a)   the temperature T for which kT is 1 electron volt
    (b)   the value of kT in electron volts for room temperature.

8.4. Calculate (a) the most probable, (b) the mean and (c) the root mean square velocity for a Maxwell velocity distribution.

8.5. In the Estermann, Simpson and Stern experiment, Cs-atoms were used and the oven temperature was $450°K$. If each of the distances from oven to slit and from slit to wire was $1·00$ m, show that the most probable distance for atoms to have fallen is 0.174 mm. [Atomic weight of Cs is 133.]

8.6. By considering the work done by an expanding gas, when no heat enters or leaves the system, show that the pressure and volume of the gas are related through

$$pV^\gamma = \text{constant}$$

under these conditions, which are known as *adiabatic*. [Hint: The work done equals the decrease in the total energy of the gas, and the total energy is given by (8.55).]

8.7. For an elastic compression of a gas, the stress is given by the change in pressure, dp, and the strain by the proportional change in volume, $-dV/V$. Show that the modulus of elasticity is given by p for an isothermal compression and by γp for an adiabatic compression.

Solutions

1. (C) Work is done by the gas in expanding at constant pressure. Hence additional heat energy must be supplied.

2. (C) At very low temperatures certain degrees of freedom are 'frozen out'.

Revision Test

1. The magnitude of the acceleration of a particle moving with uniform speed v in a circle of radius r with angular velocity ω is

   (A) $\dfrac{v}{r}$

   (B) $r\omega$

   (C) $v\omega$

   (D) vr

   (E) $\dfrac{\omega}{r}$

2. A particle coming from $x = -\infty$ with a given energy meets a potential region and is first retarded and then accelerated, ending up at $x = +\infty$.

   (A) The region is a well and the energy greater than the depth of the well.
   (B) The region is a well and the energy less than the depth of the well.
   (C) The region is a barrier and the energy greater than the height of the barrier.
   (D) The region is a barrier and the energy less than the height of the barrier.
   (E) The region is a barrier and the energy may be greater or less than the height of the barrier.

3. Two particles P_1 and P_2 move with velocities $\mathbf{v}_1$ and $\mathbf{v}_2$ respectively. If $v_1 > 2v_2$, the velocity $\mathbf{v}_{12}$ of P_1 relative to P_2 is such that

   (A) $v_{12} > v_1$

   (B) $v_{12} < v_1$

   (C) $v_{12} > v_2$

   (D) $v_{12} < v_2$

   (E) None of these.

4. Two inertial observers, moving with a small constant velocity relative to each other, make measurements on a particle acted on by a force. Their measurements will be the same for

    (A) the mass and momentum of the particle
    (B) the mass of the particle and the force acting on it
    (C) the kinetic energy of the particle
    (D) the momentum of the particle and the force acting on it
    (E) none of these.

5. In an experiment, the speeds of two electrons moving in opposite directions are found to be $0.5\,c$ and $0.8\,c$ respectively where c is the velocity of light. Their relative speed, to one decimal place, is

    (A) $0.3\,c$
    (B) $0.7\,c$
    (C) $0.9\,c$
    (D) $1.1\,c$
    (E) $1.3\,c$

6. Which one of the following expressions gives the torque about the origin of a particle of mass m with position vector $\mathbf{r}$?

    (A) $\mathbf{r} \times m\mathbf{r}$
    (B) $\mathbf{r} \times m\dot{\mathbf{r}}$
    (C) $\mathbf{r} \times m\ddot{\mathbf{r}}$
    (D) $\dot{\mathbf{r}} \times m\dot{\mathbf{r}}$
    (E) $\dot{\mathbf{r}} \times m\ddot{\mathbf{r}}$

7. A satellite of mass m is orbiting the earth in a circle of radius R. If the mass of the earth is M, the total energy of the satellite is

    (A) $-\dfrac{GmM}{R}$

    (B) $-\dfrac{GmM}{2R}$

    (C) 0

    (D) $\dfrac{GmM}{2R}$

    (E) $\dfrac{GmM}{R}$

*8.   Before   After

In the above elastic collision, which one of the following statements is incorrect?

(a)   If $v_1 > 0$, then $m_1 > m_2$

(B)   If $v_1 < v_2$, then $m_1 < m_2$

(C)   If $v_1 = 0$, then $m_1 = m_2$

(D)   If $v_2 = 3v_1$, then $m_1 = 3m_2$

(E)   If $v_2 = -v_1$, then $m_2 = 3m_1$

9. A nuclear particle disintegrates spontaneously in flight into two fragments. Which one of the following is not conserved?

(A)   The total energy in the CM system

(B)   The total energy in the LAB system

(C)   The total momentum in the CM system

(D)   The total momentum in the LAB system

(E)   The total mass

10. Three particles, of masses $1, 2$ and 2 units respectively, are located at the points $(5, 0, -5), (4, 3, 2), (1, 2, 3)$. The position of their centre of mass is

(A) $(2, 1, 0)$

(B) $(3, 2, 1)$

(C) $(3, 2, 3)$

(D) $(0, 10, 6)$

(E) $(15, 10, 5)$

11. A sphere of mass M, radius R and moment of inertia $\frac{2}{5}MR^2$ about a diameter, is rolled up an incline, making an angle α with the horizontal. If it starts with angular velocity ω, and comes to rest after travelling a distance s, the work done in that distance is

(A) $\frac{1}{5}MR^2\,\omega$

(B) $\frac{1}{5}MR^2\,\omega^2$

(C) $\frac{2}{5}MR^2\,\omega^2$

(D) Mgs

(E) $Mgs\cos\alpha$.

12. In an incompressible fluid, the velocity along a stream tube v and the cross sectional area of the tube A are such that

   (A) $v \propto A$

   (B) $v \propto A^{-\frac{1}{2}}$

   (C) $v = \text{constant}$

   (D) $v \propto A^{-\frac{1}{2}}$

   (E) $v \propto A^{-1}$

13. If the temperature of an ideal gas is raised from $27°C$ to $127°C$, the average kinetic energy of a molecule of the gas is multiplied by

   (A) $\frac{127}{27}$

   (B) $\sqrt{\frac{127}{27}}$

   (C) $\frac{4}{3}$

   (D) $\sqrt{\frac{4}{3}}$

   (E) None of these.

14. For monatomic gases, the specific heat at constant volume is

   (A) inversely proportional to the atomic weight
   (B) proportional to the temperature of the gas
   (C) proportional to the average molecular speed
   (D) greater than for diatomic gases
   (E) the same for all monatomic gases

15. The specific heat of a solid at very low temperatures is

   (A) always greater than at room temperature
   (B) always less than at room temperature
   (C) sometimes greater and sometimes less than at room temperature
   (D) independent of temperature
   (E) zero

16. A particle of unit mass moves along a straight line under a force $F = -5s^3$, where $s(t)$ is the distance from the origin at time t. The motion is fully determined if we know

   (A) $s(0)$ only, but not if we know $\dot{s}(0)$ only
   (B) $\dot{s}(0)$ only, but not if we know $s(0)$ only
   (C) either $s(0)$ or $\dot{s}(0)$
   (D) both $s(0)$ and $\dot{s}(0)$
   (E) something additional to $s(0)$ and $\dot{s}(0)$

17. The work done in stretching a spring of natural length 0·20 m to 0·25 m is 0·50 J. The work done in stretching it a further 0·05 m is

   (A)   0·50 J
   (B)   1·00 J
   (C)   1·50 J
   (D)   2·00 J
   (E)   some other value

18. A man in a lift experiences a decrease in weight. From this he can conclude that he is

   (A)   moving upwards
   (B)   moving downwards
   (C)   accelerating
   (D)   retarding
   (E)   none of these

19. The half-life of a radioactive particle, when at rest in the laboratory, is 10 s. If the particle moves at half the speed of light relative to the laboratory, then to an observer at rest in the laboratory the half-life appears to be

   (A)   7·5 s
   (B)   8·7 s
   (C)   11·5 s
   (D)   13·3 s
   (E)   20·0 s

20. A particle of mass m has a velocity 0·8c. Its kinetic energy in units of mc^2 is

   (A)   0·32
   (B)   0·50
   (C)   0·67
   (D)   1·32
   (E)   1·67

21. A particle is moving under a central force. Of the kinetic energy T, potential energy V, total energy E, linear momentum $\mathbf{p}$ and angular momentum $\mathbf{l}$, the following remain constant throughout the motion:

   (A) $E, \mathbf{p}$
   (B) $E, \mathbf{l}$
   (C) $T, \mathbf{p}$
   (D) $T, \mathbf{l}$
   (E) T, V

22. If a particle is moving under a central force $F = -\dfrac{dV}{dr}$ with angular momentum l and total energy E, then the radial equation of motion is

$$\tfrac{1}{2}m\dot{r}^2 + \frac{l^2}{2mr^2} + V(r) = E.$$

For a bound state, we have for all r that

(A) $E < V$

(B) $E > V$

(C) $E < V + \dfrac{l^2}{2mr^2}$

(D) $E > V + \dfrac{l^2}{2mr^2}$

(E)   none of these

*23. A particle mass m, velocity v, meets an absorbing material of density ρ. The particles of the absorber have mass M and the cross-section for the scattering of the incident particles by one of the particles in the absorber is σ. The mean free path of the incident particle in the absorbing material depends on

(A) m, M, σ
(B) m, ρ, σ
(C) M, ρ, σ
(D) v, ρ, σ
(E) M, v, σ

*24. A circular cylinder, that has its centre of mass a distance b from its axis, rolls or slides down an inclined plane. It is permissible to use the equation $\dot{\mathbf{L}} = \mathbf{N}$ for moments about the line of contact

(A)   if the plane is perfectly rough, but not if it is smooth
(B)   only if the plane is perfectly rough and also $b = 0$
(C)   if the plane is perfectly smooth, but not if it is rough
(D)   only if the plane is perfectly smooth and also $b = 0$
(E)   if the plane is perfectly rough or perfectly smooth and also $b = 0$

25. If the angular velocity of a body rotating about an axis is doubled, and the moment of inertia is halved, the rotational kinetic energy is

(A)   quadrupled
(B)   doubled
(C)   unchanged
(D)   halved
(E)   quartered

26. In the expression

$$m\ddot{\mathbf{r}} = m\mathbf{f} + m\omega^2\mathbf{r} - 2m\omega \times \dot{\mathbf{r}}$$

for the acceleration $\ddot{\mathbf{r}}$ in a rotating frame of reference,

(A) $m\mathbf{f}$ is an inertial force
(B) $m\omega^2\mathbf{r}$ is the centripetal force directed towards the axis of rotation
(C) $2m\boldsymbol{\omega} \times \dot{\mathbf{r}}$ is a force dependent on the position of the particle
(D) $2m\boldsymbol{\omega} \times \dot{\mathbf{r}}$ is a force parallel to the axis of rotation
(E) none of the above is correct

27. The free surface of the liquid in a beaker is found to be horizontal. We can conclude that the beaker is not

(A) being pushed along a table
(B) sliding down an inclined plane
(C) in a lift accelerating upwards
(D) on a dining car table of a train that is slowing down
(E) moving in any manner

28. The lift of an aircraft is due to the fact that

(A) the air flow is faster over the top of the wing than below it
(B) the air flow is slower over the top of the wing than below it
(C) the air flow is directed downwards by the tilt of the wing
(D) the shape of the wings creates an air cushion which supports the aircraft
(E) the velocity of the aircraft greatly exceeds the wind velocity

29. If k and N_A are Bolzmann's and Avogadro's constants, and if C_p and C_V are the specific heats at temperature T of a kilomole of a gas at constant pressure and volume respectively, then the number of degrees of freedom of a molecule of the gas can be determined from a knowledge of

(A) C_p, k, T
(B) C_p, k, N_A
(C) C_V, k, T
(D) C_V, N_A, T
(E) C_p, N_A, T

30-33. A particle increases its velocity from $\frac{3}{5}c$ to $\frac{4}{5}c$. Questions 30-33 ask for the ratio of the value of a certain property of the particle after the increase to the value before the increase. Which of the following lettered choices describes this in each case?

(A) $\frac{4}{3}$

(B) 1

(C) $\frac{16}{9}$

(D) $\frac{8}{3}$

(E) $\frac{16}{3}$

30. Total energy

31. Kinetic energy

32. Momentum

33. Rest mass

34-38. A wheel, consisting of a uniform circular disc of mass M and radius R, is mounted on a horizontal axle of radius b. A string is wound round the axle and has a weight W fixed to its free end. The weight is released and drops through distance h to a point P. The acceleration due to gravity is g. If the moment of inertia of the axle can be neglected, which of the following lettered choices give you the constants required to answer each of the questions 34-38?

(A) M, R
(B) M, R, b
(C) W, b
(D) M, R, W, b
(E) M, R, W, b, h, g

34. The moment of inertia of the wheel about the centre line of the axle.

35. The torque of the weight about the centre line of the axle.

36. The angular velocity of the wheel, when the weight is at P.

37. The angular acceleration of the wheel, when the weight is at P.

38. The kinetic energy of the weight, when it is at P.

39–42. Which of the following graphs represents best the answers to questions 39–42?

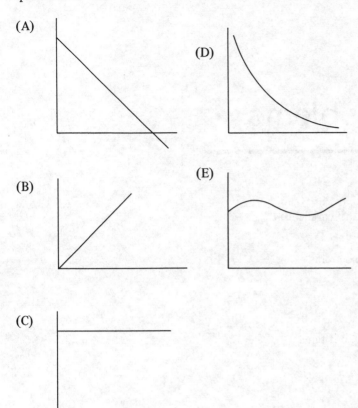

(A)

(D)

(B)

(E)

(C)

39. The speed–time graph of a planet in an elliptic orbit.

40. The graph of potential energy against distance of separation of two like electric charges.

41. The acceleration–time graph for a projectile launched upwards at an angle of $60°$ to the horizontal.

42. The velocity–time graph for a projectile launched vertically upwards.

Solutions to problems

Chapter 1

1.1 (a) $\mathbf{A} \parallel \mathbf{B}$, (b) $\mathbf{B} = 0$, (c) $\mathbf{A} \perp \mathbf{B}$, (d) $\mathbf{A}$ anti-parallel to $\mathbf{B}$.

1.2. (b) 108 km, (c) 5·46 min.

1.3. 0·024 ms^{-2} downward at 51° 30′ to the radius.

1.4. (a) At right angles to the river
(b) Upstream at 53° to the river bank.

1.5. 28 min.

1.6. 1·8 N. The average is over the distance, not time.

1.7. $\sqrt{g/l}$.

1.9. (a) $l/\sqrt{2}$, (b) $l\sqrt{K/m}$.

1.10. (a) $\dfrac{2xV_0}{(x^2 + a^2)^2}$, (b) repulsive everywhere, (c) 0.

Chapter 2

2.2 (a) $\dfrac{F}{m + M} - g$, (b) $\dfrac{Fm}{m + M}$, (c) $\sqrt{\dfrac{2h}{g}}$, (d) $(M + m)g$

2.3. Path difference $\simeq 1·5 \times 10^{-7}$ m.

2.5. 0·98c.

2.8. The difference is 1 per cent at 0·11c and 10 per cent at 0·3c.

2.9. (a) $\frac{2}{5}M$, (b) $\frac{1}{10}Mc^2$.

Chapter 3

3.1. 35 800 km. The path of the satellite must lie in a plane through the earth centre.

3.2. $11\cdot2$ km s^{-1}, $2\cdot4$ km s^{-1}.

3.6. (a) Attractive for $r^6 > \dfrac{2B}{A}$ (b) $-\dfrac{A^2}{4B}$ (c) $\left(\dfrac{B}{A}\right)^{1/6}$

3.8. $T = \pi \sqrt{\dfrac{R}{g}}$.

Chapter 4

4.5. $3\cdot9\ mc^2$, $12\cdot3\ mc^2$.

4.6. (a) $V \to c$, (b) $V \to \frac{1}{2}v_1$.

4.7. $r^2 = x^2 + y^2 + z^2$.

$$\theta = \tan^{-1}\left(\pm\frac{\sqrt{x^2 + y^2}}{z}\right), \quad 0 \leqslant \theta \leqslant \pi$$

$$\varphi = \tan^{-1}\frac{y}{x}, \quad 0 \leqslant \varphi < \pi \quad \text{for } y > 0,\ \pi < \varphi < 2\pi \quad \text{for } y < 0.$$

(a) $(\sqrt{3}, 54° \ 45', 45°)$, (b) $(\sqrt{2}, 90°, 315°)$, (c) $(\sqrt{2}, 135°, 0°)$.
(d) $(1, 0, \text{indeterminate})$.

4.9. $\dfrac{d\sigma}{d\Omega} = a^2 \operatorname{cosec}^4 \dfrac{\theta}{2}$. It is the same as Rutherford scattering.

4.10. $\sigma \simeq 0\cdot3 \times 10^{-26}$ m^2
As the radius of an atom is $\sim 10^{-10}$ m, this is evidence that the atom is largely empty.

Chapter 5

5.1. $(\frac{13}{6}, -\frac{1}{2}, \frac{2}{3})$.

5.5. $N_S/N_M = 0\cdot45$.

Chapter 6

6.1. (a) 1, (b) 5, (c) 6.

6.3. (a) $N = \dfrac{Mr^2}{2t}(\omega_0 - \omega_1)$

(b) $N = \dfrac{Mr^2 \omega}{2t} \ln \dfrac{\omega_0}{\omega_1}$.

6.5. (a) $\frac{7}{5}R$ above the table.

(b) The ball slips backwards at the point of contact. Hence the frictional force acts forwards and causes an acceleration.

6.6. (a) When the radius vector makes an angle $\cos^{-1}\frac{2}{3}$ with the vertical.

(b) The angle becomes $\cos^{-1}\frac{10}{17}$.

6.7. $2H$.

6.8. $5g/2R\omega_1$.

6.11. (a) 427 kgf, (b) 56·6 kgf.

Chapter 7

7.1. (a) The surface is inclined at an angle $\tan^{-1}(f/g)$ to the horizontal.
(b) The surface remains horizontal.

7.4. The vertical through the centre of gravity of the displaced liquid always goes through the centre of the hemisphere, which is therefore the metacentre.

7.7. $2\cdot2$ m s^{-1}.

Chapter 8

8.2. (a) 1900 m s^{-1}, 475 m s^{-1}. (b) 360 m s^{-1}, 470 m s^{-1}, 980 m s^{-1}.

8.3. (a) 11 600 K, (b) 0·026 eV.

8.4. (a) $\sqrt{\dfrac{2kT}{m}}$, (b) $\sqrt{\dfrac{8kT}{\pi m}}$, (c) $\sqrt{\dfrac{3kT}{m}}$.

Revision Test

1.	D	8.	B	15.	B	22.	D	29.	B	36.	E
2.	C	9.	E	16.	D	23.	C	30.	A	37.	D
3.	C	10.	B	17.	C	24.	B	31.	D	38.	E
4.	B	11.	B	18.	E	25.	B	32.	C	39.	E
5.	C	12.	E	19.	C	26.	E	33.	B	40.	D
6.	C	13.	C	20.	C	27.	D	34.	A	41.	C
7.	B	14.	E	21.	B	28.	A	35.	C	42.	A

Index

Printed by William Clowes & Sons Limited, London, Colchester and Beccles